High Collection Nonimaging Optics

High Collection Nonimaging Optics

W. T. WELFORD

Optics Section
Department of Physics
Imperial College of Science, Technology and Medicine
University of London
London, England

R. WINSTON

Enrico Fermi Institute and
Department of Physics
University of Chicago
Chicago, Illinois

ACADEMIC PRESS, INC.
Harcourt Brace Jovanovich, Publishers
San Diego New York Berkeley Boston
London Sydney Tokyo Toronto

Cover photograph: Computer graphics display of rays traced through the secondary. Courtesy of Philip Gleckman.

This book is printed on acid-free paper. ♾

Academic Press, Inc.
San Diego, California 92101

United Kingdom Edition published by
Academic Press Limited
24–28 Oval Road, London NW1 7DX

Library of Congress Cataloging-in-Publication Data

Welford, W. T.
High collection nonimaging optics / W. T. Welford, R. Winston.
p. cm.
Includes index.
ISBN 0-12-742885-2 (alk. paper)
1. Optical instruments. I. Winston, Roland. II. Title.
III. Title: Nonimaging optics.
TS514.W45 1989
621.36--dc19 89-322
CIP

Printed in the United States of America
89 90 91 92 9 8 7 6 5 4 3 2 1

Contents

Chapter 11 **Illumination of Optical Systems and Instruments**

Appendix A **Derivation and Explanation of the Étendue Invariant, Including the Dynamical Analogy; Derivation of the Skew Invariant**

Preface

Many developments have occurred in the field of nonimaging optics since we published the first book on the subject in 1978. The emphasis then was on applications to concentrators for solar energy, although in fact the ideas originated in connection with instrumentation for high energy physics. Applications to solar energy concentration are now accepted and widespread, but since 1978 developments have occurred in two directions. One of these is in the development of new formalisms in nonimaging optics, in particular in the use of the geometrical vector flux concept, and these have led to entirely new concentrator designs. The other direction is in wider applications in nonsolar work; nonimaging components now find uses in a wide range of instruments, both for collecting radiation and redirecting it from a source.

In this book we have tried to cover these new areas while at the same time retaining the coverage of pre-1978 material. The result is a comprehensive text on all aspects of nonimaging optics. The title reflects the present-day wider range of applications of the subject.

The contributions and encouragement of our colleagues and students have facilitated continued progress in this field. We are particularly grateful to Oscar Manley of the United States Department of Energy, Office of Basic Energy Sciences for steadfastly supporting our collaboration. We thank Susan Hubbard for editing and typing the manuscript. Finally, we thank Jill Goldenberg for adding "High Collection" to the title.

W. T. Welford
R. Winston

CHAPTER 1

Concentrators and Their Uses

1.1 Concentrating Collectors

Nonimaging collectors have several actual and some potential applications, but it is convenient to explain the general concept of a nonimaging concentrator by reference to the most important application, the utilization of solar energy. The radiation power density received from the sun at the earth's surface, often denoted by S, peaks at approximately 1 kW m^{-2}, depending on many factors. If we attempt to collect this power by absorbing it on a perfect blackbody, the equilibrium temperature T of the blackbody will be given by*

$$\sigma T^4 = S \tag{1.1}$$

where σ is the Stefan–Boltzmann constant, 5.67×10^{-8} W m^{-2} °K^{-4}.

In this example the equilibrium temperature would be 364°K, or just below the boiling point of water.

For many practical applications of solar energy this is adequate, and it is well known that systems for domestic hot-water heating based on this principle are available commercially for installation in private dwellings. However, for larger-scale purposes or for generating electric

* Ignoring various factors such as convection and conduction losses and radiation at lower effective emissivities.

power a source of heat at 364°K has a low thermodynamic efficiency since it is not practicable to get a very large temperature difference in whatever working fluid is being used in the heat engine. If we wanted, say, $\geq 300°C$, a useful temperature for the generation of motive power, we should need to increase the power density S on the absorbing blackbody by a factor C of about 6–10 from Eq. (1.1).

This, briefly, is one use of a concentrator—to increase the power density of solar radiation. Stated baldly like that the problem sounds trivial. The principles of the solution have been known since the days of Archimedes and his burning glass*: We simply have to focus the image of the sun with an image-forming system, i.e., a lens, and the result will be an increased power density. The problems to be solved are technical and practical, but they lead also to some interesting pure geometrical optics. The first question is that of the maximum concentration, i.e., how large a value of C is theoretically possible? This question has simple answers in all cases of interest. The next question, Can the theoretical maximum concentration be achieved in practice? is not so easy to answer. We shall see that there are limitations involving materials and manufacturing, as we should expect. But also there are limitations involving the kinds of optical systems that can actually be designed, as opposed to those that are theoretically possible. This is analogous to the situation in classical lens design. The designers sometimes find that a certain specification cannot be fulfilled because it would require an impractically large number of refracting or reflecting surfaces. But sometimes also they do not know whether it is in principle possible to achieve aberration corrections of a certain kind.

The natural approach of the classical optical physicist is to regard the problem as one of designing an image-forming optical system of very large numerical aperture, i.e., small aperture ratio or *f*-number. One of the most interesting results to have emerged in this field is that there is a class of very efficient concentrators that would have very large aberrations if they were used as image-forming systems. Nevertheless, as concentrators they are substantially more efficient than image-forming systems and can be designed to meet or approach the theoretical limit. We shall call them nonimaging concentrating collectors, or nonimaging concentrators for short. These systems are very unlike any previously used optical systems. They have some of

* For an amusing argument concerning the authenticity of the story of Archimedes, see Stavroudis (1973).

the properties of light pipes and some of the properties of image-forming optical systems, but with very large aberrations. The development of the designs of these concentrators and the study of their properties have led to a range of new ideas and theorems in geometrical optics. In order to facilitate the development of these ideas it is necessary to recapitulate some basic principles of geometrical optics, which is done in Chapter 2. In Chapter 3 we look at what can be done with conventional image-forming systems as concentrators and we show how they necessarily fall short of ideal performance. In Chapter 4 we describe one of the basic nonimaging concentrators, the compound parabolic concentrator, and we obtain its optical properties. Chapter 5 is devoted to several developments of the basic compound parabolic concentrator with plane absorber, mainly aimed at decreasing the overall length. In Chapter 6 we examine in more detail concentrators with troughlike geometry in which the surface is cylindrical (but not a circular cylinder). These have particular applications to solar energy concentration with absorbers such as tubes of nonplane shape. In Chapter 7 we introduce further theoretical developments in geometrical optics, and in Chapter 8 wave optics aspects are discussed. Chapter 9 discusses the accuracy with which the shapes of concentrators must be made, and Chapter 10 is devoted specifically to systems intended for solar energy concentration. Finally, in Chapter 11 we examine briefly several applications unconnected with solar energy. There are several appendixes in which the derivations of the more complicated formulas are given.

1.2 Definition of the Concentration Ratio; the Theoretical Maximum

From the simple argument in Section 1.1 we see that the most important property of a concentrator is the ratio of area of input beam divided by area of output beam; this is because the equilibrium temperature of the absorbing body is proportional to the fourth root of this ratio. We denote this ratio by C and call it the *concentration ratio.* Initially we model a concentrator as a box with a plane entrance aperture of area A and a plane exit aperture of area A' that is just large enough to allow all transmitted rays to emerge (Fig. 1.1). Then

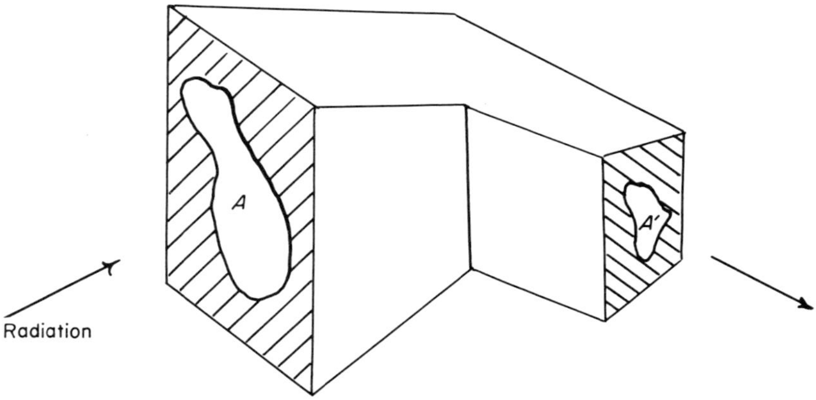

Fig. 1.1 Schematic diagram of a concentrator. The input and output surfaces can face in any direction; they are drawn as above so that both can be seen. It is assumed that the aperture A' is just large enough to permit all rays passed by the internal optics that have entered within the specified collecting angle to emerge.

the concentration ratio is

$$C = A/A' \tag{1.2}$$

In the above definition it was tacitly assumed that compression of the input beam occurred in both the dimensions transverse to the beam direction, as in ordinary lens systems. In solar energy technology there is a large class of systems in which the beam is compressed in only one dimension. In such systems all the operative surfaces, reflecting and refracting, are cylindrical with parallel generators (but not in general circular cylindrical). Thus a typical shape would be as in Fig. 1.2, with the absorbing body, not shown, lying along the trough. Such long trough collectors have the obvious advantage that they do not need to be guided to follow the daily movement of the sun across the sky. The two types of concentrators are sometimes called three- and two-dimensional, or 3D and 2D, concentrators. The concentration ratio of a 2D concentrator is usually given as the ratio of the transverse input and output dimensions, measured perpendicular to the straight line generators of the trough.

The question immediately arises whether there is any upper limit to the value of C, and we shall see that there is. The result, to be proved later, is very simple for the 2D case and for the 3D case with

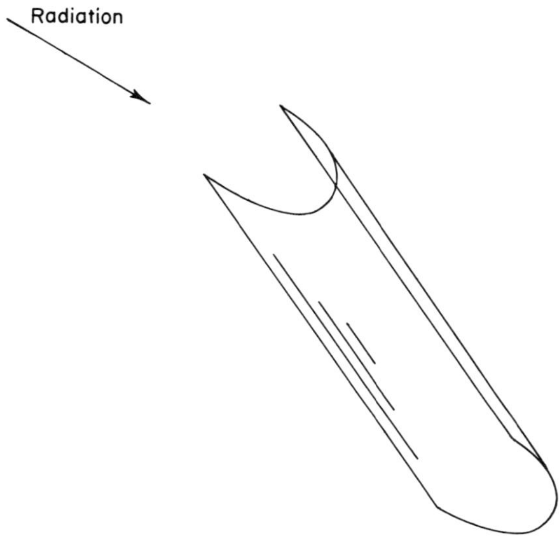

Fig. 1.2 A trough concentrator; the absorbing element is not shown.

an axis of revolution symmetry. Suppose the input and output media both have a refractive index of unity, and let the incoming radiation be from a circular source at infinity subtending a semiangle θ_i. Then the theoretical maximum concentration in 3D is

$$C_{\max} = 1/\sin^2 \theta_i \qquad (1.3)$$

Under this condition the rays emerge at all angles up to $\pi/2$ from the normal to the exit face, as in Fig. 1.3. For a 2D system the corresponding value will be seen to be $1/\sin \theta_i$.

The next question that arises is, Can actual concentrators be designed with the theoretically best performance? In asking this question we make certain idealizing assumptions, e.g., all reflecting surfaces have 100% reflectivity, all refracting surfaces can be perfectly antireflection coated, all shapes can be made exactly right, etc. We shall then see that the following answers are obtained. (1) 2D collectors can be designed with the theoretical maximum concentration. (2) 3D collectors can also have the theoretical maximum concentration if they have spherical symmetry in their construction. (3) Some 3D collectors with cylindrical symmetry (i.e., an axis of revolution) can have the theoretical maximum concentration. In case (3) it appears for other types of

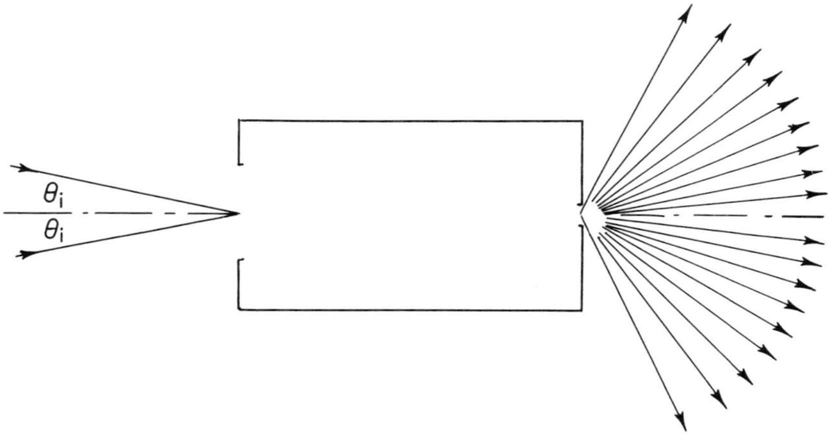

Fig. 1.3 Incident and emergent ray paths for an ideal 3D concentrator with symmetry about an axis of revolution. The exit aperture diameter is sin θ_i times the entrance aperture diameter; the rays emerge from all points in the exit aperture over a solid angle 2π.

design that it is possible to approach indefinitely close to the theoretical maximum concentration either by sufficiently increasing the complexity of the design or by incorporating materials that are in principle possible but in practice not available. For example, we might specify a material of very high refractive index, say, 5, although this is not actually available without large absorption in the visible part of the spectrum.

1.3 Uses of Concentrators

The application to solar energy utilization mentioned above has, of course, stimulated the greatest developments in the design and fabrication of concentrators. But this is by no means the only application. The particular kind of nonimaging concentrator that has given rise to the greatest developments was originally conceived as a device for collecting as much light as possible from a luminous volume (the gas or fluid of a Čerenkov counter) over a certain range of solid angle and sending it onto the cathode of a photomultiplier. Since photomultipliers are limited in size and the volume in question was of order

1 m^3, this is clearly a concentrator problem (Hinterberger and Winston, 1966a,b).

Subsequently the concept was applied to infrared detection (Harper *et al.*, 1976), where it is well known that the noise in the system for a given type of detector increases with the surface area of the detector (other things being equal).

Another application of a different kind was to the optics of visual receptors. It has been noted (Winston and Enoch, 1971) that the cone receptors in the human retina have a shape corresponding approximately to that of a nonimaging concentrator designed for approximately the collecting angle that the pupil of the eye would subtend at the retina under dark-adapted conditions.

There are several other possible applications of nonimaging concentrators, and these will be discussed in Chapter 11.

CHAPTER 2

Some Basic Ideas in Geometrical Optics

2.1 The Concepts of Geometrical Optics

Geometrical optics is used as the basic tool in designing almost any optical system, image forming or not. We use the intuitive ideas of a ray of light, roughly defined as the path along which light energy travels, together with surfaces that reflect or transmit the light. When light is reflected from a smooth surface it obeys the well-known law of reflection, which states that the incident and reflected rays make equal angles with the normal to the surface and that both rays and the normal lie in one plane. When light is transmitted, the ray direction is changed according to the law of refraction, Snell's law. This law states that the sine of the angle between the normal and the incident ray bears a constant ratio to the sine of the angle between the normal and the refracted ray; again, all three directions are coplanar.

A major part of the design and analysis of concentrators involves ray tracing, i.e., following the paths of rays through a system of reflecting and refracting surfaces. This is a well-known process in conventional lens design but the requirements are somewhat different for concentrators, so it will be convenient to state and develop the methods ab initio. In conventional lens design the reflecting or refracting surfaces involved are almost always portions of spheres, and the centers of the spheres lie on one straight line (axisymmetric optical system), so that special methods that take advantage of the simplicity of the forms of the surfaces and the symmetry can be used. Nonimaging concentrators do not, in general, have spherical surfaces. In fact, sometimes there is no explicit analytical form for the surfaces, although usually there is an axis or a plane of symmetry. We shall find it most convenient, therefore, to develop ray-tracing schemes based on vector formulations, but with the details dealt with in computer programs on an ad hoc basis for each different shape.

In geometrical optics we represent the power density across a surface by the density of ray intersections with the surface, and the total power by the number of rays. This notion, reminiscent of the useful but outmoded "lines of force" in electrostatics, works as follows. We take N rays spaced uniformly over the entrance aperture of a concentrator at an angle of incidence θ, as in Fig. 2.1. Suppose that after tracing the rays through the system only N' emerge through the exit aperture, the dimensions of the latter being determined by the desired concentration ratio. The remaining $N - N'$ rays are lost by processes that will become clear when we consider some examples. Then the power transmission for the angle θ is taken as N'/N. This can be extended to cover a range of angles θ as required. Clearly, N must be taken large enough to ensure that a thorough exploration of possible ray paths in the concentrator is made.

2.2 Formulation of the Ray-Tracing Procedure

In order to formulate a ray-tracing procedure suitable for all cases, it is convenient to put the laws of reflection and refraction into vector

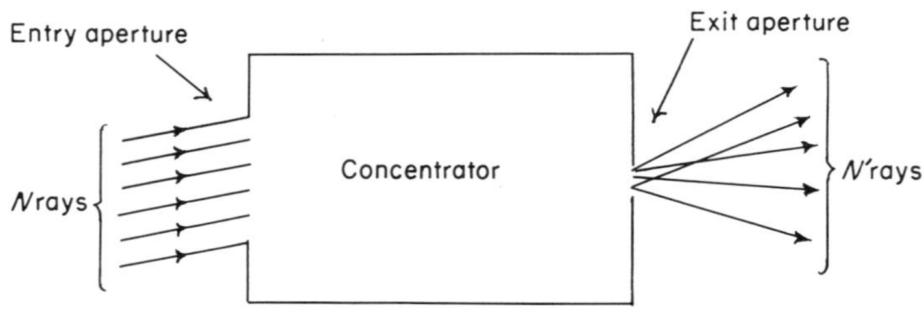

Fig. 2.1 Determining the transmission of a concentrator by ray tracing.

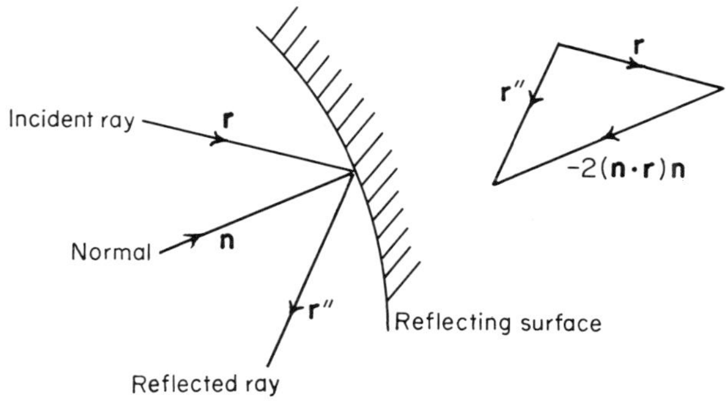

Fig. 2.2 Vector formulation of reflection. **r**, **r**″, and **n** are all unit vectors.

form. Figure 2.2 shows the geometry with unit vectors **r** and **r**″ along the incident and reflected rays and a unit vector **n** along the normal pointing into the reflecting surface. Then it is easily verified that the law of reflection is expressed by the vector equation

$$\mathbf{r}'' = \mathbf{r} - 2(\mathbf{n} \cdot \mathbf{r})\mathbf{n} \tag{2.1}$$

as in the diagram.

Thus to ray trace "through" a reflecting surface, first we have to find the point of incidence, a problem of geometry involving the direction of the incoming ray and the known shape of the surface. Then we have to find the normal at the point of incidence, again a problem of geometry. Finally, we have to apply Eq. (2.1) to find the direction of the reflected ray. The process is then repeated if another reflection is to be taken into account. These stages are illustrated in Fig. 2.3. Naturally, in the numerical computation the unit vectors are represented by their components, i.e., the direction cosines of the ray or normal with respect to some cartesian coordinate system used to define the shape of the reflecting surface.

Ray tracing through a refracting surface is similar, but first we have to formulate the law of refraction vectorially. Figure 2.4 shows the relevant unit vectors. It is similar to Fig. 2.2 except that **r**′ is a unit vector along the refracted ray. We denote by n, n' the refractive indices of the media on either side of the refracting boundary; the refractive index is a parameter of a transparent medium related to the speed of light in the medium. Specifically, if c is the speed of light in vacuum, the speed in a transparent material medium is c/n, where n is the

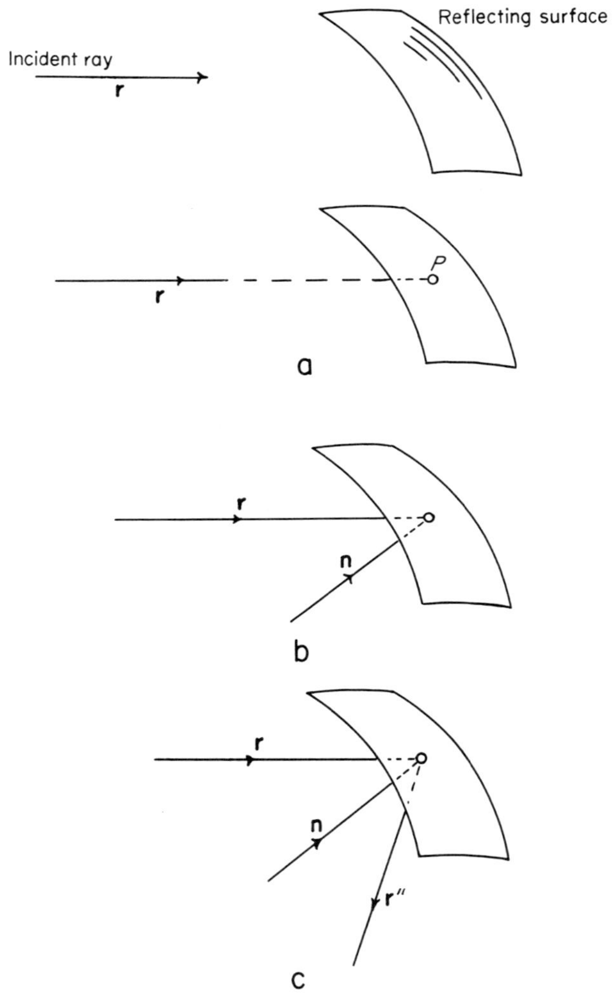

Fig. 2.3 The stages in ray tracing a reflection. (a) Find the point of incidence P. (b) Find the normal at P. (c) Apply Eq. (2.1) to find the reflected ray $\mathbf{r}''$.

refractive index. For visible light, values of n range from unity to about 3 for usable materials in the visible spectrum. The law of refraction is usually stated in the form

$$n' \sin I' = n \sin I \tag{2.2}$$

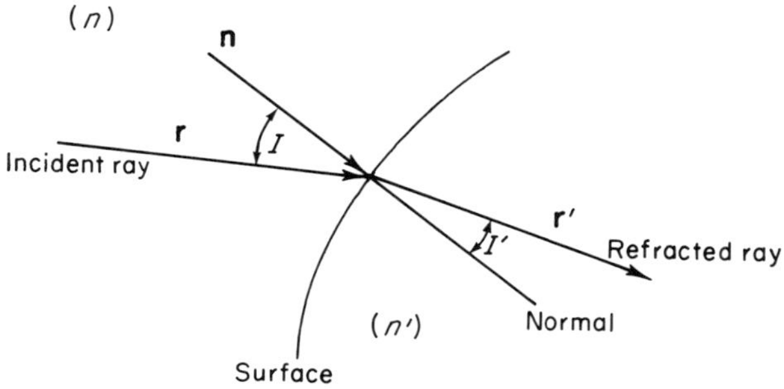

Fig. 2.4 Vector formulation of refraction.

where I and I' are the angles of incidence and refraction, as in Fig. 2.4, and where the coplanarity of the rays and the normal is understood. The vector formulation

$$n'\mathbf{r}' \times \mathbf{n} = n\mathbf{r} \times \mathbf{n} \tag{2.3}$$

contains everything, since the modulus of a vector product of two unit vectors is the sine of the angle between them. This can be put in the form most useful for ray tracing by multiplying through vectorially by **n** to give

$$n'\mathbf{r}' = n\mathbf{r} + (n'\mathbf{r}' \cdot \mathbf{n} - n\mathbf{r} \cdot \mathbf{n})\mathbf{n} \tag{2.4}$$

which is the preferred form for ray tracing.* The complete procedure then parallels that for reflection explained by means of Fig. 2.3. We find the point of incidence, then the direction of the normal, and finally the direction of the refracted ray. Details of the application to lens systems are given, e.g., by Welford (1974, 1986).

If a ray travels from a medium of refractive index n toward a boundary with another of index $n' < n$, then it can be seen from Eq. (2.2) that it would be possible to have $\sin I'$ greater than unity. Under this condition it is found that the ray is completely reflected at the boundary. This is called *total internal reflection* and we shall find it a useful effect in concentrator design.

* The method of using Eq. (2.4) numerically is not so obvious as for Eq. (2.2) since the coefficient of $\boldsymbol{n}$ in Eq. (2.4) is actually $n' \cos I' - n \cos I$. Thus it might appear that we have to find $\boldsymbol{r}'$ before we can use the equation. The procedure is to find $\cos I'$ via Eq. (2.2) first, and then Eq. (2.4) is needed to give the complete three-dimensional picture of the refracted ray.

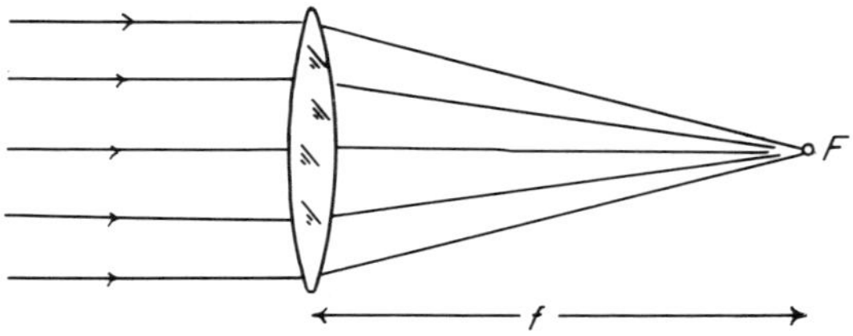

Fig. 2.5 A thin converging lens bringing parallel rays to a focus. Since the lens is technically "thin," we do not have to specify the exact plane in the lens from which the focal length f is measured.

2.3 Elementary Properties of Image-Forming Optical Systems

In principle, the use of ray tracing tells us all there is to know about the geometrical optics of a given optical system, image forming or not. However, ray tracing alone is of little use for inventing new systems having properties suitable for a given purpose. We need to have ways of describing the properties of optical systems in terms of general performance, such as, for example, the concentration ratio C introduced in Chapter 1. In this section we shall introduce some of these concepts.

Consider first a thin converging lens such as would be used as a magnifier or might be found in the spectacles of a farsighted person. In section, this looks like Fig. 2.5. By the term "thin" we mean that its thickness can be neglected for the purposes under discussion. Elementary experiments show us that if we have rays coming from a point at a great distance to the left, so that they are substantially parallel as in the figure, the rays meet approximately at a point F, the focus. The distance from the lens to F is called the focal length, denoted by f. Elementary experiments also show that if the rays come from an object of finite size at a great distance, the rays from each point on the object converge to a separate focal point and we get an image. This is, of course, what happens when a burning glass forms an image of the sun or when the lens in a camera forms an image on film. This is indicated in Fig. 2.6, where the object subtends the (small) angle 2θ. It is then found that the size of the image is $2f\theta$. This is easily seen by considering the rays through the center of the lens, since these pass through undeviated.

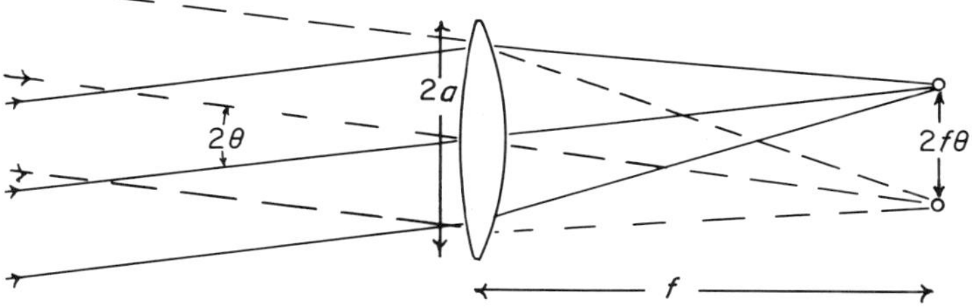

Fig. 2.6 An object at infinity has an angular subtense 2θ. A lens of focal length f forms an image of size $2f\theta$.

Figure 2.6 contains one of the fundamental concepts we use in concentrator theory, the concept of a beam of light of a certain diameter and angular extent. The diameter is that of the lens, $2a$ say, and the angular extent is given by 2θ. These two can be combined as a product, usually without the factor 4, giving θa, a quantity known by various names, such as *extent, étendue, acceptance, Lagrange invariant,* etc. It is, in fact, an invariant through the optical system, provided that there are no obstructions in the light beam and provided we ignore certain losses due to properties of the materials, such as absorption and scattering. For example, at the plane of the image the étendue becomes the image height θf multiplied by the convergence angle a/f of the image-forming rays, giving again θa. In discussing 3D systems, e.g., an ordinary lens such as we have supposed Fig. 2.6 to represent, it is convenient to deal with the square of this quantity, $a^2\theta^2$. This is also sometimes called the étendue, but generally it is clear from the context and from dimensional considerations which form is intended. The 3D form has an interpretation that is fundamental to the theme of this book. Suppose we put an aperture of diameter $2f\theta$ at the focus of the lens, as in Fig. 2.7. Then this system will only accept rays within the angular range $\pm\theta$ and inside the diameter $2a$. Now suppose a flux of radiation B W m^{-2} sr^{-1} is incident on the lens from the left.* The system will actually accept a total flux $B\,\pi^2\theta^2a^2$ W; thus the étendue or acceptance θ^2a^2 is a measure of the power flow that can pass through the system.

The same discussion shows how the concentration ratio C appears in the context of classical optics. The accepted power $B\,\pi^2\theta^2a^2$ W must flow out of the aperture to the right of the system if our

* In full, B watts per square meter per steradian solid angle.

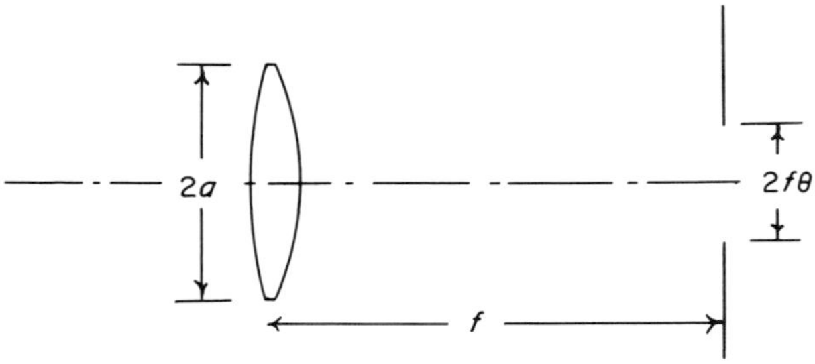

Fig. 2.7 An optical system of acceptance, throughput, or étendue $a^2\theta^2$.

assumptions above about how the lens forms an image are correct* and if the aperture has the diameter $2f\theta$. Thus our system is acting as a concentrator with concentration ratio $C=(2a/2f\theta)^2=(a/f\theta)^2$ for the input semiangle θ.

Let us relate these ideas to practical cases. For solar energy collection we have a source at infinity that subtends a semiangle of approximately 0.005 rad ($\frac{1}{4}°$), so that this is the given value of θ, the collection angle. Clearly for a given diameter of lens we gain by reducing the focal length as much as possible.

2.4 Aberrations in Image-Forming Optical Systems

According to the simplified picture presented in Section 2.3, there is no reason why we could not make a lens system with indefinitely large concentration ratio by simply decreasing the focal length sufficiently. This is, of course, not so, partly because of aberrations in the optical system and partly because of the fundamental limit on concentration stated in Section 1.2.

We can explain the concept of aberrations by reference again to our example of the thin lens of Fig. 2.5. We suggested that the parallel rays shown all converged after passing through the lens to a single point *F*. In fact, this is only true in the limiting case when the diameter of the lens is taken as indefinitely small. The theory of optical systems under this condition is called *paraxial optics* or *Gaussian optics* and it is a very useful approximation for getting at the main large-scale

* As we shall see, these assumptions are only valid for limitingly small apertures and objects.

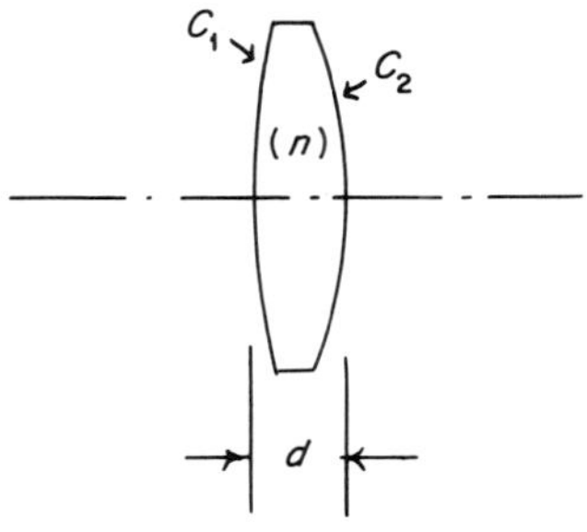

Fig. 2.8 Specification of a singlet lens. The curvature c_1 is positive as shown, and c_2 is negative.

properties of image-forming systems. If we take a simple lens with a diameter that is a sizable fraction of the focal length, say $f/4$, we find that the rays from a single point object do not all converge to a single image point. We can show this by ray tracing. We first set up a proposed lens design as in Fig. 2.8. The lens has curvatures (reciprocals of radii) c_1 and c_2, center thickness *d*, and refractive index *n*. If we neglect the central thickness for the moment, then it is shown in specialized treatments (e.g., Welford, 1986) that the focal length f is given in paraxial approximation by

$$1/f = (n-1)(c_1 - c_2) \tag{2.5}$$

and we can use this to get the system to have roughly the required paraxial properties.

Now we can trace rays through the system as specified, using the method outlined in Section 2.2 [details of ray-tracing methods for ordinary lens systems are given in, e.g., Welford (1974)]. These will be exact or finite rays, as opposed to paraxial rays, which are implicit in the Gaussian optics approximation. The results for the lens in Fig. 2.8 would look as in Fig. 2.9. This shows rays traced from an object point on the axis at infinity, i.e., rays parallel to the axis.

In general, for a convex lens the rays from the outer part of the lens aperture meet the axis closer to the lens than the paraxial rays. This effect is known as *spherical aberration.* (The term is misleading, since the aberration can occur in systems with nonspherical refracting surfaces, but there seems little point in trying to change it at the present advanced state of the subject.)

Spherical aberration is perhaps the simplest of the different aberration types to describe, but it is just one of many. Even if we were to choose the shapes of the lens surfaces to eliminate the spherical

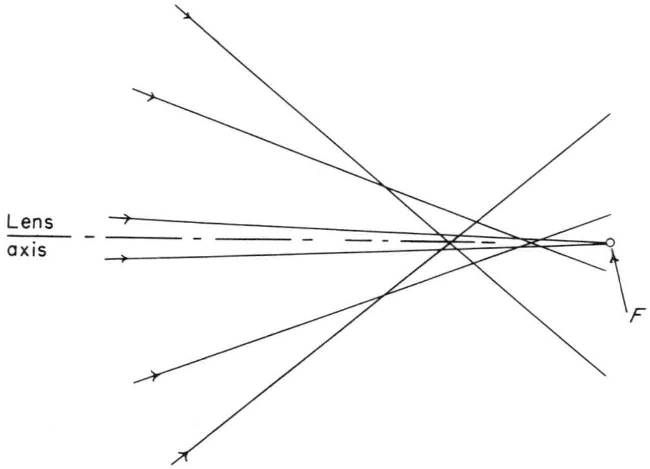

Fig. 2.9 Rays near the focus of a lens showing spherical aberration.

aberration or were to eliminate it in some other way, we should still find that the rays from object points away from the axis did not form point images, i.e., there would be oblique or off-axis aberrations. Also, the refractive index of any material medium changes with the wavelength of the light, and this produces *chromatic aberrations* of various kinds. We do not at this stage need to go into the classification of aberrations very deeply, but this preliminary sketch is necessary to show the relevance of aberrations to the attainable concentration ratio.

2.5 The Effect of Aberrations in an Image-Forming System on the Concentration Ratio

Questions regarding the extent to which it is theoretically possible to eliminate aberrations from an image-forming system have not yet been fully answered. In this book we shall attempt to give answers adequate for our purposes, although they may not be what the classical lens designers want. For the moment, let us accept that it is possible to eliminate spherical aberration completely but not the off-axis aberrations, and let us suppose that this has been done for the simple collector of Fig. 2.7. The effect will be that some rays of the beam at the extreme angle θ will fall outside the defining aperture of diameter $2f\theta$. We can

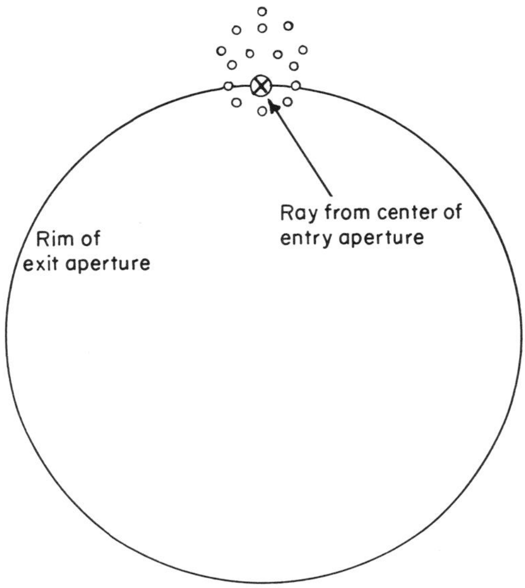

Fig. 2.10 A spot diagram for rays from the beam at the maximum entry angle for an image-forming concentrator. Some rays miss the edge of the exit aperture due to aberrations, and the concentration is thus less than the theoretical maximum.

see this more clearly by representing an aberration by means of a *spot diagram.* This is a diagram in the image plane with points plotted to represent the intersections of the various rays in the incoming beam. Such a spot diagram for the extreme angle θ might appear as in Fig. 2.10. The ray through the center of the lens (the *principal* ray in lens theory) meets the rim of the collecting aperture by definition, and thus a considerable amount of the flux does not get through. Conversely, it can be seen, in this case at least, that some flux from beams at a larger angle than θ will be collected.

We display this information on a graph such as Fig. 2.11. This shows the proportion of light collected at different angles up to the theoretical maximum, θ_{max}. An ideal collector would behave according to the full line, i.e., it would collect all light flux within θ_{max} and none outside. At this point it may be objected to that all we need do to achieve the first requirement is to enlarge the collecting aperture slightly, and that the second requirement does not matter. However, we recall that our aim is to achieve maximum concentration because of the requirement

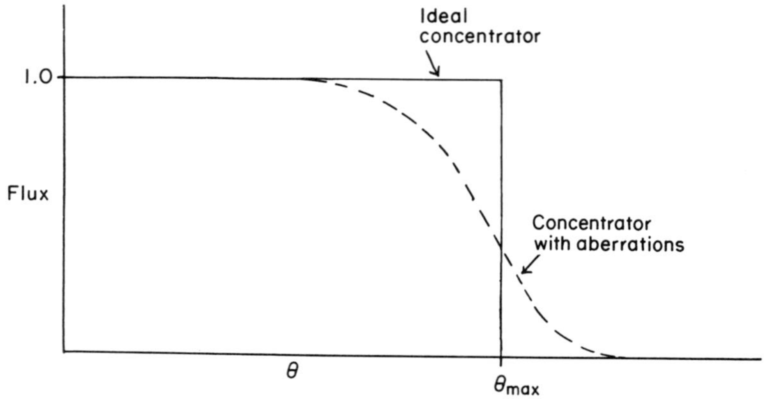

Fig. 2.11 A plot of collection efficiency against angle. The ordinate is the proportion of flux entering the collector aperture at angle θ that emerges from the exit aperture.

for high operating temperature, so that the collector aperture must not be enlarged beyond $2f\theta$ diameter.

Frequently in discussions of aberrations in books on geometrical optics the impression is given that aberrations are in some sense "small." This is true in optical systems designed and made to form reasonably good images, e.g., camera lenses. But these systems do not operate with large enough convergence angles (a/f in the notation for Fig. 2.6) to approach the maximum theoretical concentration ratio. If we were to try to use a conventional image-forming system under such conditions, we should find that the aberrations would be very large and that they would severely depress the concentration ratio. Roughly, we can say that this is one limitation that has led to the development of the new nonimaging concentrators.

2.6 The Optical Path Length and Fermat's Principle

There is another way of looking at geometrical optics and the performance of optical systems, which we also need to outline for the purposes of this book. We noted in Section 2.2 that the speed of light in a medium of refractive index n is c/n, where c is the speed in vacuum. Thus, light travels a distance s in the medium in time $s/v = ns/c$, i.e., the time taken to travel a distance s in a medium of refractive

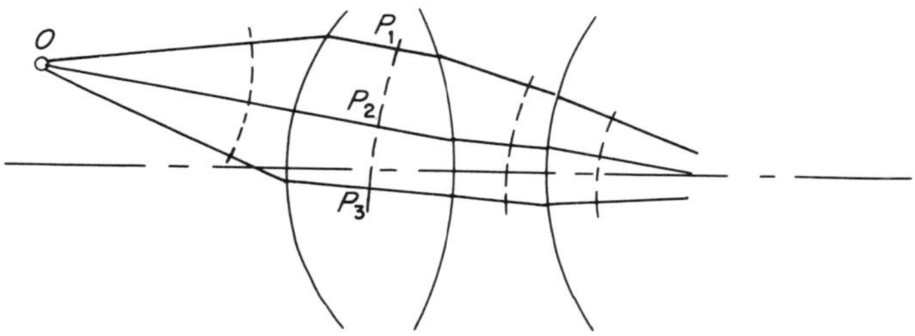

Fig. 2.12 Rays and (in broken line) geometrical wave fronts.

index n is proportional to ns. The quantity ns is called the *optical path length* corresponding to the length s. Suppose we have a point source O emitting light into an optical system, as in Fig. 2.12. We can trace any number of rays through the system, as outlined in Section 2.2, and then we can mark off along these rays points that are all at the same optical path length from O, say $P_1, P_2, \ldots$. We do this by making the sum of the optical path lengths from O in each medium the same, i.e.,

$$\sum ns = \text{const.} \tag{2.6}$$

in an obvious notation. These points can be joined to form a surface (we are supposing rays out of the plane of the diagram to be included), which would be a surface of constant phase of the light waves if we were thinking in terms of the wave theory of light.* We call it a geometrical wave front, or simply a wave front, and we can construct wave fronts at all distances along the bundle of rays from O.

We now introduce a principle that is not so intuitive as the laws of reflection and refraction but that leads to results that are indispensable to the development of the theme of this book. It is based on the concept of optical path length and it is a way of predicting the path of a ray through an optical medium. Suppose we have any optical medium which can have lenses and mirrors and can even have regions of continuously varying refractive index. We wish to predict the path of a light ray between two points in this medium, say A and B in Fig. 2.13. We can propose an infinite number of possible paths, of which three are indicated. But unless A and B happen to be object and image—and we assume they are not—only one or perhaps a small

* This construction does not give a surface of constant phase near a focus or near an edge of an opaque obstacle, but this does not affect the present applications.

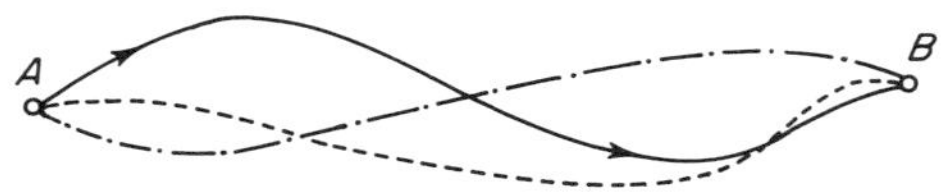

Fig. 2.13 Fermat's principle. It is assumed in the diagram that the medium has a continuously varying refractive index. The solid line path has a stationary optical path length from A to B and is therefore a physically possible ray path.

finite number of paths will be physically possible, i.e., paths that rays of light could take according to the laws of geometrical optics. Fermat's principle in the form usually used states that a physically possible ray path is one for which the optical path length along it from A to B is an extremum as compared to neighboring paths. For "extremum" we can often write "minimum," as in Fermat's original statement. It is possible to derive all of geometrical optics, i.e., the laws of refraction and reflection, from Fermat's principle. It also leads to the result that the geometrical wave fronts are orthogonal to the rays (the theorem of Malus and Dupin), i.e., the rays are normals to the wave fronts. This in turn tells us that if there is no aberration, i.e., all rays meet at one point, then the wave fronts must be portions of spheres. Thus, also, if there is no aberration, the optical path length from object point to image point is the same along all rays. Thus we arrive at an alternative way of expressing aberrations, i.e., in terms of the departure of wave fronts from the ideal spherical shape. This concept will be useful when we come to discuss the different senses in which an image-forming system can form "perfect" images.

2.7 The Generalized Étendue or Lagrange Invariant and the Phase Space Concept

We next have to introduce a concept that is essential to the development of the principles of nonimaging concentrators. In Section 2.3 we noted that there is a quantity $a^2\theta^2$ that is a measure of the power accepted by the system, where a is the radius of the entrance aperture and θ is the semiangle of the beams accepted. In paraxial approximation for an axisymmetric system this is invariant through the optical system; we considered only the regions near the entrance and exit apertures, but it is shown in specialized texts on optics that the same quantity can be written down for any region inside a complex optical

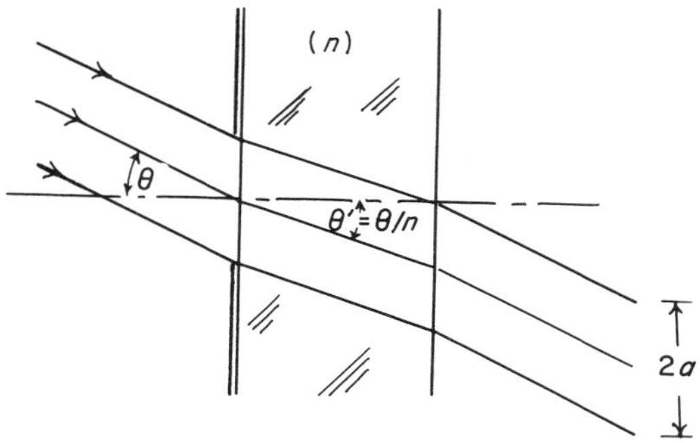

Fig. 2.14 Inside a medium of refractive index n the étendue becomes $n^2a^2\theta'^2$.

system. There is one slight complication—if we are considering a region of refractive index different from unity, say the inside of a lens or a prism, the invariant is written $n^2a^2\theta^2$. The reason for this can be seen from Fig. 2.14, which shows a beam at the extreme angle θ entering a plane-parallel plate of glass of refractive index n. Inside the glass the angle is $\theta' = \theta/n$, by the law of refraction,* so that the invariant in this region is

$$\text{étendue} = n^2a^2\theta'^2 \tag{2.7}$$

We might try to use the étendue to obtain an upper limit for the concentration ratio of a system as follows. We suppose we have an axisymmetric optical system of any number of components, i.e., not necessarily the simple system sketched in Fig. 2.7. The system will have an entrance aperture of radius a, which may be the rim of the front lens or, as in Fig. 2.15, possibly some limiting aperture inside the system. An incoming parallel beam may emerge parallel, as indicated in the figure, or not, and this will not affect the result. But to simplify the argument it is easier to imagine a parallel beam emerging from an aperture of radius a'. The concentration ratio is by definition $(a/a')^2$ and if we use the étendue invariant and assume that the initial and final media are both air or vacuum, i.e., refractive index unity, the concentration ratio becomes $(\theta'/\theta)^2$. Since from obvious geometrical considerations θ' cannot exceed $\pi/2$ this suggests $(\pi/2\theta)^2$ as a theoretical upper limit to the concentration.

* The paraxial approximation is implied, so that $\sin\theta \sim \theta$.

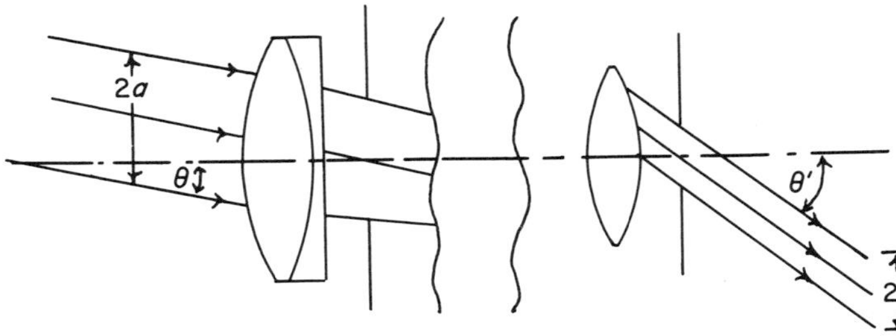

Fig. 2.15 The étendue for a multielement optical system with an internal aperture stop.

Unfortunately, this argument is invalid because the étendue as we have defined it is essentially a paraxial quantity. Thus it is not necessarily an invariant for angles as large as $\pi/2$. In fact, the effect of aberrations in the optical system is to ensure that the paraxial étendue is *not* an invariant outside the paraxial region, so that we have not found the correct upper limit to the concentration.

There is, as it turns out, a suitable generalization of the étendue to rays at finite angles to the axis and we now proceed to explain this. The concept has been known for some time but as it has not been used to any extent in classical optical design, it is not described in many texts. It applies to optical systems of any or no symmetry and of any structure—refracting, reflecting, or with continuously varying refractive index.

Let the system be bounded by homogeneous media of refractive indices n and n' as in Fig. 2.16, and suppose we have a ray traced

Fig. 2.16 The generalized étendue.

exactly between the points P and P' in the respective input and output media. We wish to consider the effect of small displacements of P and of small changes in direction of the ray segment through P on the emergent ray, so that these changes define a beam of rays of a certain cross section and angular extent. In order to do this we set up a cartesian coordinate system $Oxyz$ in the input medium and another, $O'x'y'z'$, in the output medium. The positions of the origins of these coordinate systems and the directions of their axes are quite arbitrary with respect to each other, to the directions of the ray segments, and, of course, to the optical system. We specify the input ray segment by the coordinates of $P(x, y, z)$, and by the direction cosines of the ray (L, M, N). The output segment is similarly specified. We can now represent small displacements of P by increments dx and dy to its x and y coordinates and we can represent small changes in the direction of the ray by increments dL and dM to the direction cosines for the x and y axes. Thus we have generated a beam of area $dx\,dy$ and angular extent defined by $dL\,dM$. This is indicated in Fig. 2.17 for the y section.* Corresponding increments dx', dy', dL', and dM' will occur in the output ray position and direction.

Then the invariant quantity turns out to be $n^2\,dx\,dy\,dL\,dM$, i.e., we have

$$n'^2\,dx'\,dy'\,dL'\,dM' = n^2\,dx\,dy\,dL\,dM \tag{2.8}$$

The proof of this theorem depends on other concepts in geometrical optics that we do not need in this book. We have therefore given a

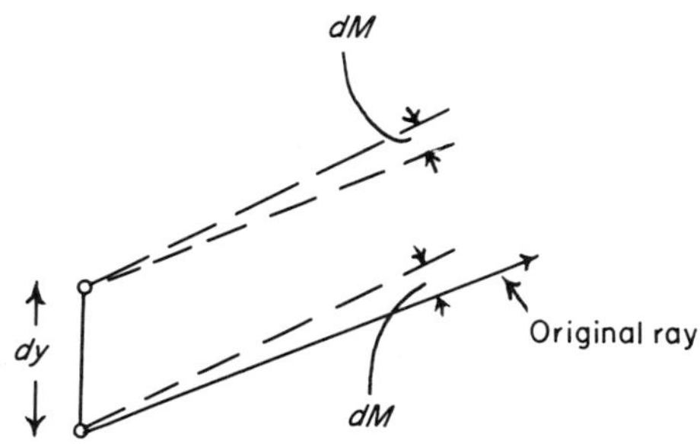

Fig. 2.17 The generalized étendue in the y section.

---

* It is necessary to note that the increments dL and dM are in direction cosines, not angles. Thus in Fig. 2.17 the notation on the figure should be taken to mean not that dM is the angle indicated, but merely that it is a measure of this angle.

proof in Appendix A, where references to other proofs of it can also be found.

The physical meaning of Eq. (2.8) is that it gives the changes in the rays of a beam of a certain size and angular extent as it passes through the system. If there are apertures in the input medium that produce this limited étendue and if there are no apertures elsewhere to cut off the beam, then the accepted light power emerges in the output medium, so that the étendue as defined is a correct measure of the power transmitted along the beam. It may seem at first remarkable that the choice of origin and direction of the coordinate systems is quite arbitrary.* However, it is not very difficult to show that the generalized étendue or Lagrange invariant as calculated in one medium is independent of coordinate translations and rotations. This, of course, must be so if it is to be a meaningful physical quantity.

The generalized étendue is sometimes written in terms of the optical direction cosines $p = nL$, $q = nM$, when it takes the form

$$dx\, dy\, dp\, dq \tag{2.9}$$

We can now use the étendue invariant to calculate the theoretical maximum concentration ratios of concentrators. Consider first a 2D concentrator as in Fig. 2.18. From Eq. (2.8) we have for any ray bundle

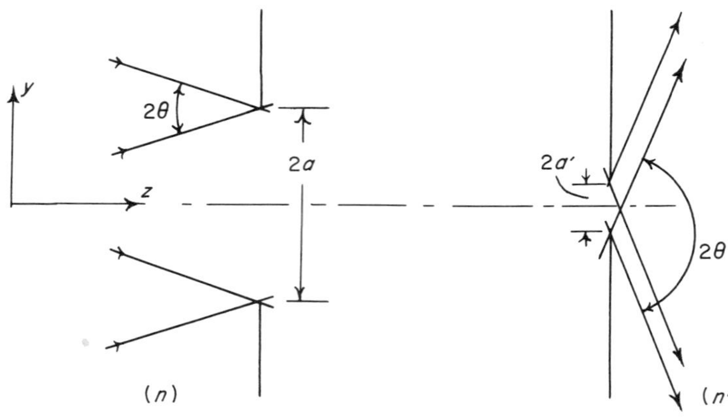

Fig. 2.18 The theoretical maximum concentration ratio for a 2D optical system.

---

* This is not quite true. It can be seen from the formulation of the theorem that we cannot choose a direction for the z axis that is perpendicular to the ray direction.

that traverses the system

$$n\,dy\,dM = n'\,dy'\,dM' \tag{2.10}$$

and integrating over y and M we obtain

$$4na \sin\theta = 4n'a' \sin\theta' \tag{2.11}$$

so that the concentration ratio is

$$a/a' = (n' \sin\theta')/(n \sin\theta) \tag{2.12}$$

In this result a' is a dimension of the exit aperture large enough to permit any ray that reaches it to pass, and θ' is the largest angle of all the emergent rays. Clearly, θ' cannot exceed $\pi/2$, so the theoretical maximum concentration ratio is

$$C_{\max} = n'/(n \sin\theta) \tag{2.13}$$

Similarly, for the 3D case we can show that for an axisymmetric concentrator the theoretical maximum is

$$C_{\max} = (a/a')^2 = [n'/(n \sin\theta)]^2, \tag{2.14}$$

where again θ is the input semiangle.

The results in Eqs. (2.13) and (2.14) are maximum values, which may or may not be attained. We may find in practice that if the exit aperture has the diameter given by Eq. (2.14), some of the rays within the incident collecting angle and aperture do not pass it. We sometimes also find in a number of the systems to be described that some of the incident rays are actually turned back by internal reflections and never reach the exit aperture. In addition, there are losses due to absorption, imperfect reflectivity, etc., but these do not represent fundamental limitations. Thus Eqs. (2.13) and (2.14) give theoretical upper bounds on performance of concentrators.

Our results so far apply to 2D concentrators [Eq. (2.13)] with rectangular entrance and exit apertures and to 3D axisymmetric concentrators with circular entrance and exit apertures [Eq. (2.14)]. We ought, for completeness, to discuss briefly what happens if the entrance aperture is not circular but the concentrator itself still has an axis of symmetry. The difficulty with this case is that it depends on the details of the internal optics of the concentrator. It may happen that the internal optical system forms an image of the entrance aperture on the exit aperture, in which case it would be correct to make them similar in shape. For an entry aperture of arbitrary shape but uniform entry angle $\pm\theta_i$, all that can be said in general is that for an ideal concentrator the area of the exit aperture must equal that of the entry aperture multiplied by $\sin^2\theta_i$; however, it does not follow that such an ideal concentrator could be designed for any arbitrary shape of entry aperture.

2.8 The Skew Invariant

There is an invariant associated with the path of a skew ray through an axisymmetric optical system. Let S be the shortest distance between the ray and the axis, i.e., the length of the common perpendicular, and let γ be the angle between the ray and the axis. Then the quantity

$$h = nS \sin\gamma \tag{2.15}$$

is an invariant through the whole system. If the medium has a continuously varying refractive index, the invariant for a ray at any coordinate z_1 along the axis is obtained by treating the tangent of the ray at the z value as the ray and using the refractive index value at the point where the ray cuts the transverse plane z_1. The skew-invariant formula will be proved in Appendix A.

If we use the dynamical analogy described in Appendix A, then h corresponds to the angular momentum of a particle following the ray path and the skew-invariant theorem corresponds to conservation of angular momentum.

2.9 Different Versions of the Concentration Ratio

We now have some different definitions of concentration ratio. It is desirable to clarify them by using different names. First, in Section 2.7 we established upper limits for the concentration ratio in 2D and 3D systems, given respectively by Eqs. (2.13) and (2.14). These upper limits depend only on the input angle and the input and output refractive indices. Clearly, we can call either expression the *theoretical maximum concentration ratio.*

Second, an actual system will have entry and exit apertures of dimensions $2a$ and $2a'$. These can be width or diameter for 2D or 3D systems, respectively. The exit aperture may or may not transmit all rays that reach it, but in any case the ratios (a/a') or $(a/a')^2$ define a *geometrical concentration ratio.*

Third, given an actual system we can trace rays through it and determine the proportion of incident rays within the collecting angle that emerge from the exit aperture. This process will yield an *optical concentration ratio.*

Finally, we could make allowances for attenuation in the concentrator by reflection losses, scattering, manufacturing errors, and absorption in calculating the optical concentration ratio. We could call the result the *optical concentration ratio with allowance for losses.*

The optical concentration ratio will always be less than or equal to the theoretical maximum concentration ratio. The geometrical concentration ratio can, of course, have any value.

CHAPTER 3

Some Designs of Image-Forming Concentrators

3.1 Introduction

In this chapter we shall study image-forming concentrators of conventional form, e.g., paraboloidal mirrors, lenses of short focal length, etc., and estimate their performance. Then we shall show how the departure from ideal performance suggests a principle for the design of nonimaging concentrators, the "edge-ray principle" as we shall call it.

3.2 Some General Properties of Ideal Image-Forming Concentrators

In order to fix our ideas we use the solar energy application to describe the mode of action of our systems. The simplest hypothetical image-forming concentrator would then function as in Fig. 3.1. The rays are coded to indicate that rays from one direction from the sun are brought to a focus at one point in the exit aperture, i.e., the concentrator images the sun (or other source) at the exit aperture. If the exit medium is air, then the exit angle θ' must be $\pi/2$ for maximum concentration. Clearly, such a concentrator would, in practice, have to

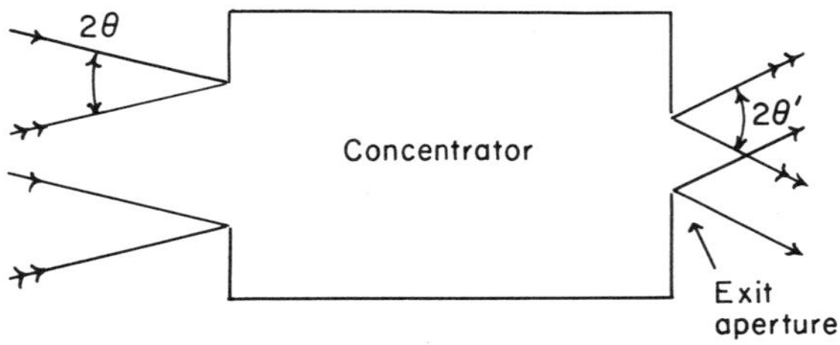

Fig. 3.1 An image-forming concentrator. An image of the source at infinity is formed at the exit aperture of the concentrator.

be constructed with glass or some other medium of refractive index greater than unity forming the exit surface, as in Fig. 3.2. Also, the angle θ' in the glass would have to be such that $\sin\theta' = 1/n$ so that the emergent rays just fill the required $\pi/2$ angle. For typical materials the angle θ' would be about 40°.

Figure 3.2 brings out an important point about the optics of such a concentrator. We have labeled the central or principal ray of the two extreme angle beams a and b, respectively, and at the exit end these rays have been drawn normal to the exit face. This would be essential if the concentrator were to be used with air as the final medium since, if rays a and b were not normal to the exit face, some of the extreme angle rays would be totally internally reflected (see Section 2.2) and thus the concentration ratio would be reduced. In fact, the condition that the exit principal rays should be normal to what, in ordinary lens design, is termed the image plane is not usually

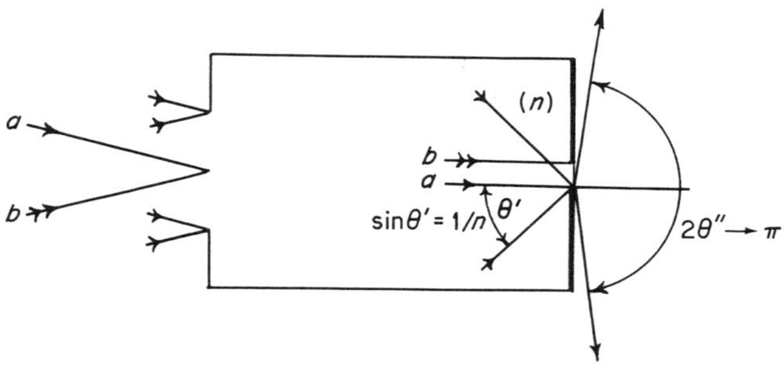

Fig. 3.2 In an image-forming concentrator of maximum theoretical concentration ratio, the final medium in the concentrator would have to have a refractive index n greater than unity. The angle θ' in this medium would be $\arcsin(1/n)$, giving an angle $\pi/2$ in the air outside.

fulfilled. Such an optical system, called *telecentric*, needs to be specially designed, and the requirement imposes constraints that would certainly worsen the attainable performance of a concentrator. We shall therefore assume that when a concentrator ends in glass of index n, the absorber or other means of utilizing the light energy is placed in optical contact with the glass in such a way as to avoid potential losses through total internal reflection.

An alternative configuration for an image-forming concentrator would be as in Fig. 3.3. The concentrator collects rays over θ_{max} as before, but the internal optics forms an image of the entrance aperture at the exit aperture, as indicated by the arrow coding of the rays. This would be in optics terminology a *telescopic* or *afocal* system. Naturally, the same considerations about using glass or a similar material as the final medium holds as for the system of Figs. 3.1 and 3.2, and there is no difference between the systems as far as external behavior is concerned.

If the concentrator terminates in a medium of refractive index n, we can gain in maximum concentration ratio by a factor n or n^2, depending on whether it is a 2D or 3D system, as can be seen from Eqs. (2.13) and (2.14). This corresponds to having an extreme angle $\theta' = \pi/2$ in this medium. We then have to reinstate the requirement that the principal rays be normal to the exit aperture, and we also have to ensure that the absorber can utilize rays of such extreme angles.

In practice there are problems in utilizing extreme collection angles approaching $\theta' = \pi/2$ whether in air or a higher index medium. There has to be very good matching at the interface between glass and absorber to avoid large reflection losses of grazing-incidence rays, and irregularities of the interface can cause losses through shadowing. Therefore we may well be content with values of θ' of, say, 60°. This

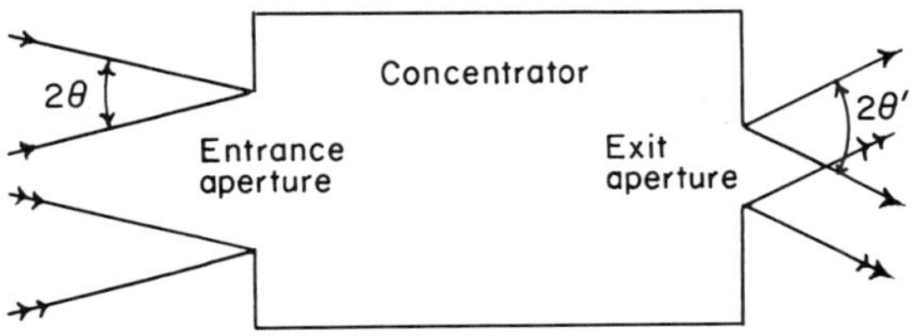

Fig. 3.3 An alternative configuration of an image-forming concentrator. The rays collected from an angle $\pm\theta$ form an image of the entrance aperture at the exit aperture.

represents only a small decrease from the theoretical maximum concentration, as can be seen from Eqs. (2.13) and (2.14).

Thus in speaking of ideal concentrators, we can also regard as ideal a system that brings all incident rays within θ_{max} out within θ'_{max} and inside an exit aperture a' given by Eq. (2.12), i.e., $a' = na \sin \theta_{max}/n' \sin \theta'_{max}$. Such a concentrator will be ideal, but will not have the theoretical maximum concentration.

The concentrators sketched in Figs. 3.1 and 3.2 clearly must contain something like a photographic objective with very large aperture (small *f*-number), or perhaps a high-power microscope objective used in reverse. The speed of a photographic objective is indicated by its *f*-number or aperture ratio. Thus an $f/4$ objective has a focal length four times the diameter of its entrance aperture. This description is not suitable for imaging systems in which the rays form large angles approaching $\pi/2$ with the optical axis for a variety of reasons. It is found that in discussing the resolving power of such systems the most useful measure of performance is the *numerical aperture*, or NA, a concept introduced by E. Abbe in connection with the resolving power of microscopes. Figure 3.4 shows an optical system with entrance aperture of diameter $2a$. It forms an image of the axial object point at infinity, and the semiangle of the cone of extreme rays is α'_{max}. Then

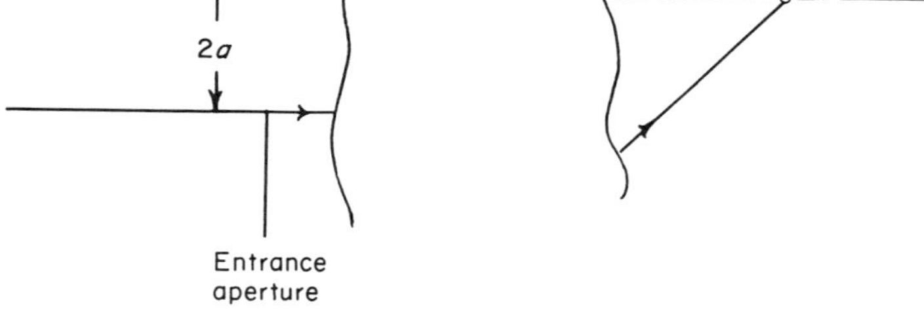

Fig. 3.4 The definition of the numerical aperture of an image forming system. The NA is $n' \sin \alpha'$.

the numerical aperture is defined by

$$\mathrm{NA} = n' \sin \alpha'_{\max} \tag{3.1}$$

where n' is the refractive index of the medium in the image space. We assume that all the rays from the axial object point focus sharply at the image point, i.e., there is (to use the terminology of Section 2.4) no spherical aberration. Then Abbe showed that off-axis object points will also be sharply imaged if the condition

$$h = n' \sin \alpha' \times \text{const.} \tag{3.2}$$

is fulfilled for all the axial rays. In this equation h is the distance from the axis of the incoming ray and α' is the angle at which that ray meets the axis in the final medium. Equation (3.2) is a form of the celebrated Abbe sine condition for good image formation. It does not ensure perfect image formation for all off-axis object points, but it ensures that aberrations that grow linearly with the off-axis angle are zero. These aberrations are various kinds of *coma* and the condition of freedom from spherical aberration and coma is called *aplanatism.*

Clearly, a necessary condition for our image-forming concentrator to have the theoretical maximum concentration—or even for it to be ideal as an image-forming system (but without theoretical maximum concentration)—is that the image formation should be aplanatic. This is not, unfortunately, a sufficient condition.

The constant in Eq. (3.2) has the significance of a focal length. The definition of focal length of optical systems with media of different refractive indices in the object and image spaces is more complicated than for the thin lenses discussed in Chapter 2. In fact, it is necessary to define two focal lengths, one for the input space and one for the output space, where their magnitudes are in the ratio of the refractive indices of the two media. In Eq. (3.2) it turns out that the constant is the input side focal length, which we shall denote by f.

From Eq. (3.2) we have for the input semiaperture

$$a = f \cdot \mathrm{NA} \tag{3.3}$$

and also, from Eq. (2.12),

$$a' = a \sin \theta_{\max} / (\mathrm{NA}) \tag{3.4}$$

By substituting from Eq. (3.3) into Eq. (3.4) we have

$$a' = f \sin \theta_{\max} \tag{3.5}$$

where θ_{max} is the input semiangle. To see the significance of this result, we recall that we showed that in an aplanatic system the focal length is a constant, independent of the distance h of the ray from the axis used to define it. Here we are using the generalized sense of "focal length" meaning the constant in Eq. (3.2), and aplanatism thus means that rays through all parts of the aperture of the system form images with the same magnification. Thus Eq. (3.5) tells us that in an imaging concentrator with maximum theoretical concentration, the diameter of the exit aperture is proportional to the sine of the input angle. This is true even if the concentrator has a numerical exit aperture less than the theoretical maximum, n', provided it is ideal in the sense defined above.

From the point of view of conventional lens optics the result of Eq. (3.3) is well known. It is simply another way of saying that the aplanatic lens with largest aperture and with air as the exit medium is $f/0.5$ since Eq. (3.5) tells us that $a = f$. The importance of Eq. (3.5) is that it tells us something about one of the shape-imaging aberrations required of the system, namely distortion. A distortion-free lens imaging onto a flat field must obviously have an image height proportional to tan θ, so that our concentrator lens system is required to have what is usually called *barrel distortion.* This is illustrated in Fig. 3.5.

Our picture of an imaging concentrator is gradually taking shape and we can begin to see that certain requirements of conventional imaging can be relaxed. Thus if we can get a sharp image at the edge of the exit aperture and if the diameter of the exit aperture fulfills the requirement of Eqs. (3.3)–(3.5), we do not need perfect image formation for object points at angles smaller than θ_{max}. For example, the image field perhaps could be curved, provided we take the exit aperture in the plane of the circle of image points for the direction θ_{max}, as in Fig. 3.6. Also, the inner parts of the field could have point-imaging aberrations, provided these were not so large as to spill rays outside the circle of radius a'. Thus we see that an image-forming concentrator need not, in principle, be so difficult to design as an imaging lens since the aberrations need to be corrected only at the edge of the field. In practice this relaxation may not be very helpful because the outer part of the field is the most difficult to correct. However, this leads us to a valuable principle for nonimaging concentrators. Not only is it unnecessary to have good aberration correction except at the exit rim, but we do not even need point imaging at the rim itself. It is only necessary that rays entering at the extreme angle θ_{max} should leave

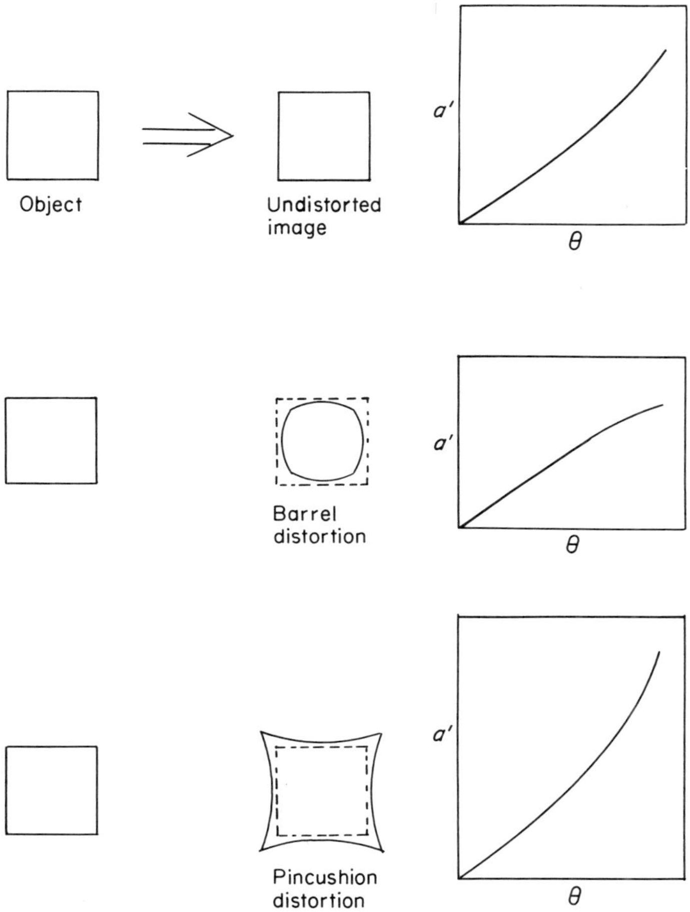

Fig. 3.5 Distortion in image-forming systems. The optical systems are assumed to have symmetry about an axis of rotation.

from some point at the rim and that the aberrations inside should not be such as to push rays outside the rim of the exit aperture. We shall return to this *edge-ray principle* later in connection with nonimaging concentrators.

The above arguments need only a little modification to apply to the alternative configuration of an imaging concentrator in Fig. 3.3, in which the entrance aperture is imaged at the exit aperture. Referring

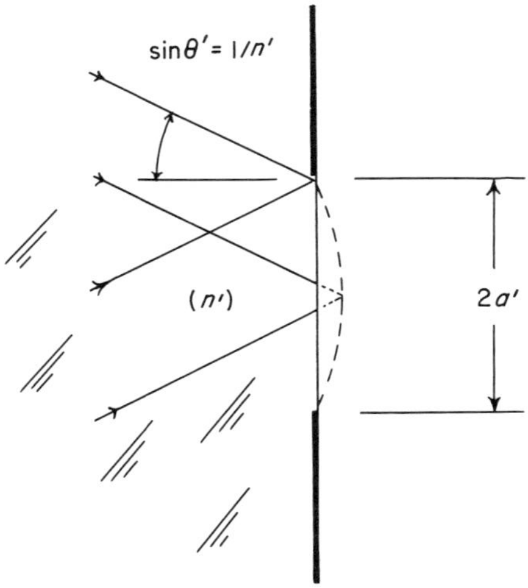

Fig. 3.6 A curved image field with a plane exit aperture.

to Fig. 3.7, we can imagine that the optical components of the concentrator are forming an image at the exit aperture of an object at a considerable distance, rather than at infinity, and that this object is the entrance aperture. Alternatively, we can imagine that part of the concentrator is a collimating lens of focal length *f*, shown in broken line in the figure, and that this projects the entrance aperture to infinity with an angle subtending $2a/f$. The same considerations as before then apply to the aberration corrections.

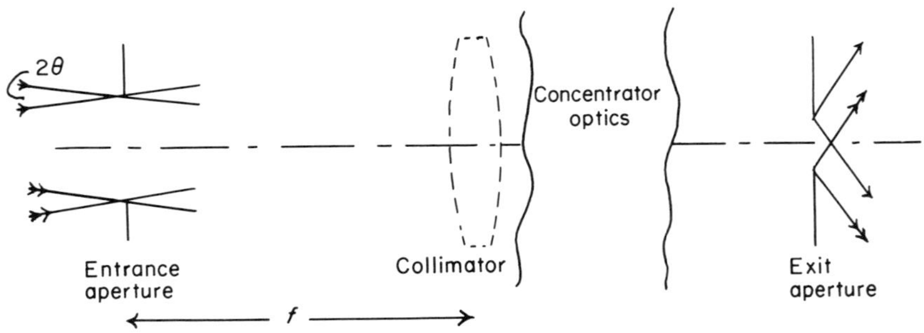

Fig. 3.7 An afocal concentrator shown as two image-forming systems.

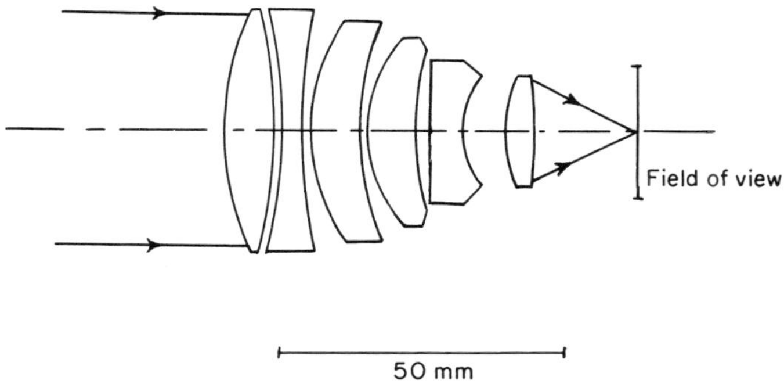

Fig. 3.8 A high-aperture camera objective. The drawing is to scale for a 50-mm focal length. The emerging cone of rays has a semiangle of 26° at the center of the field of view.

3.3 Can an Ideal Image-Forming Concentrator Be Designed?

In Section 3.2 we outlined some requirements for ideal image-forming concentrators, and we now have to ask whether they can be designed to fulfill these requirements with useful collection angles.

High-aperture camera lenses are made at about $f/1.0$ but these are complex structures with many components. Figure 3.8 shows a typical example with a focal length of 50 mm. Such a system is by no means aberration free, and the cost of scaling it up to a size useful for solar work would be prohibitive. Anyway, its numerical aperture is still only about 0.5. The only systems with numerical apertures approaching the theoretical limit are microscope objectives. Figure 3.9 shows one of the simplest designs of microscope objective of numerical aperture

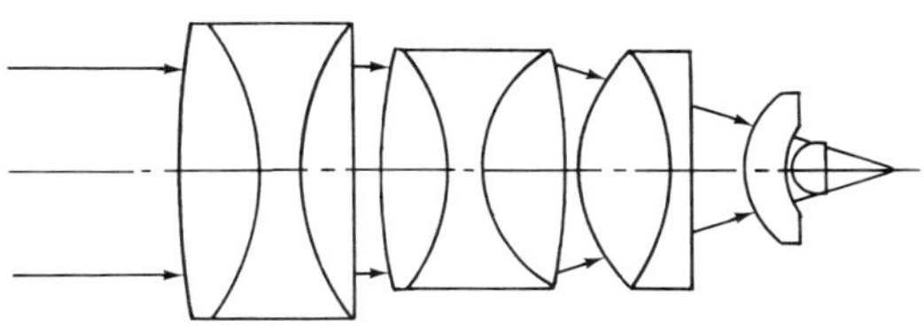

Fig. 3.9 An oil-immersion microscope objective of high numerical aperture. Such systems can have a convergence angle of up to 60° with an aberration-free field of about ±3°. However, they can only be designed aberration free for focal lengths up to 2 mm, i.e., an actual field diameter of about 200 μm.

about 1.35, drawn in reverse and with one conjugate at infinity. The image or exit space has a refractive index of 1.52 since it is an oil-immersion objective. Such systems have good aberration correction only to about 3° from the axis. Beyond this the aberrations increase rapidly and also there is less light transmission because of vignetting.* The collecting aperture would be about 4 mm in diameter. Again, it would be impracticable to scale up such a system to useful dimensions.

Thus a quick glance at the state of the art in conventional lens design suggests that imaging concentrators in the form of lens systems will not be very efficient on a practical scale. Nevertheless, it is interesting to see what might be done if practical limitations are ignored. Unfortunately, conventional lens theory is incomplete and there are certain theorems about the possibility of lens designs of various degrees of perfection that have been conjectured but never proved.

Roughly, the position seems to be that we cannot design an ideal concentrator, i.e., one with the theoretical maximum collection efficiency, using a finite number of lens elements. But, by increasing the number of elements sufficiently or by postulating sufficiently extreme optical properties, we can approach indefinitely close to the ideal. We cannot prove this rigorously, but in Appendix B we give some arguments that will be sufficiently convincing to those who have tried their hands at optical design.

Apparent exceptions to the above proposed rule occur in optical systems with spherical symmetry. It has been known since the time of Huygens that a spherical lens element images a concentric surface as in Fig. 3.10. The two conjugate surfaces have radii r/n and nr, respectively. The configuration is used in microscope objectives having a high numerical aperture, as in Fig. 3.9. Unfortunately, one of the conjugates must always be virtual (the object conjugate as the figure is drawn) so that the system alone would not be very practical as a concentrator. It seems to be true, although this has not been proven, that no combination of a finite number of concentric components can form an aberration-free real image of a real object. However, as we shall see, this can be done with media of continuously varying refractive index. The system of Fig. 3.10 would clearly be useful as the last stage of an imaging concentrator. It can easily be shown that the

* Vignetting is caused by rims of components at either end of a long system shearing against each other as the system is turned off-axis.

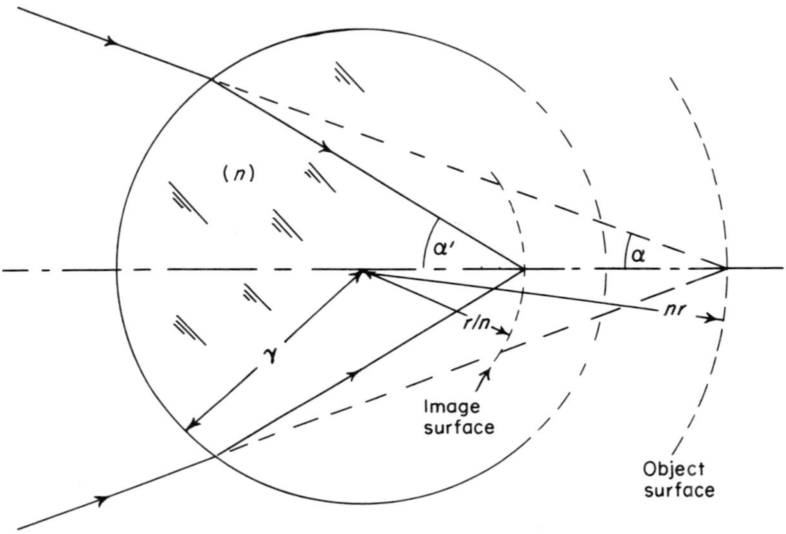

Fig. 3.10 The aplanatic surfaces of a spherical refracting surface.

convergence angles are related by the equation

$$\sin \alpha' = n \sin \alpha \tag{3.6}$$

Also, if there is a plane surface terminating in air, the final emergent angle α'' is given by

$$\sin \alpha'' = n^2 \sin \alpha \tag{3.7}$$

Thus this system could be used in conjunction with another system of relatively low numerical aperture, as in Fig. 3.11, to form a fairly well-corrected concentrator. This is, of course, merely a reinvention of the microscope objective of Fig. 3.9, and the postulated additional system still needs to operate at about $f/1$ if ordinary materials are used. If we assume some extreme material qualities, say a refractive index of 4 with adequate antireflection coating for the aplanatic component, then the auxiliary system only needs to be $f/8$ to give $\pi/2$ emergent angles in air, as in Fig. 3.12. This is not a difficult requirement. In fact, it is probably true that if we ignore chromatic aberration, i.e., if we assume our postulated material of index 4 has no dispersion, this system could be designed in moderate sizes for which the aberrations are indefinitely small over a reasonable acceptance angle.

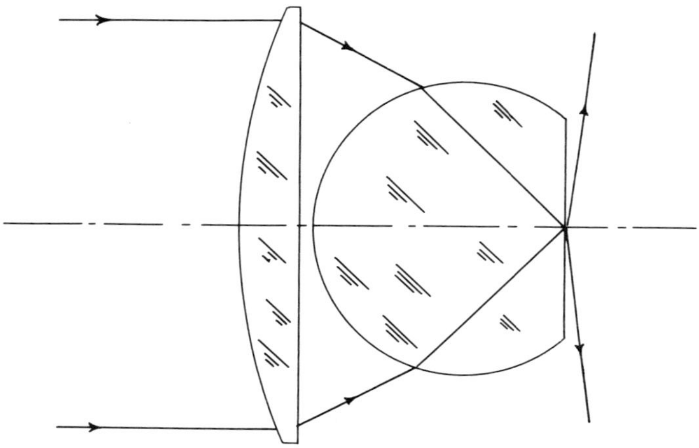

Fig. 3.11 An image-forming concentrator with an aplanatic component.

Ultimately, the ray aberrations of optical systems become negligible because the performance, as an imaging system, is limited by diffraction effects. Thus if an imaging optical system of a certain numerical aperture is used to form an image of a point source and if there are precisely no ray aberrations, the image of the point will not be indefinitely small. It is shown in books on physical optics* that the point image will be a blurred diffraction pattern in which most of the light flux falls inside a circle of radius

$$0.61\lambda / NA \tag{3.8}$$

where λ is the wavelength of the light. This provides us with a tolerance level for ray aberrations below which we can say the aberrations are

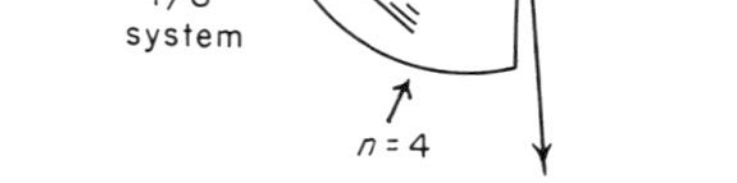

Fig. 3.12 Use of an aplanatic component of high refractive index to produce a well-corrected optical system.

* See, e.g., Born and Wolf (1975).

negligible. This is sometimes expressed in the form that all of the points in the spot diagram for the aberrations (see Fig. 2.10) must fall within a circle of radius given by Eq. (3.8). Another way of setting a tolerance is to say that the wave-front shape as determined by the methods outlined in Section 2.6 should not depart from the ideal spherical shape by more than a specified amount, usually $\lambda/4$. Other tolerance systems are also described by Born and Wolf (1975).

Thus in some way we could arrive at tolerances for the geometrical aberrations such that an imaging concentrator with aberrations inside these tolerances would have ideal performance. Any lens system made with available materials would be impractically complicated and costly if scaled up to a size suitable for solar concentration. But hypothetical materials could be used to bring the aberrations within the diffraction limit with a simple design such as in Fig. 3.12.

At this point it may be of interest to offer a different example of a lens system as a concentrator. It is well known that an ellipsoidal solid lens will focus parallel light without spherical aberration, as in Fig. 3.13, provided the eccentricity is equal to $1/n$. In order to use this as

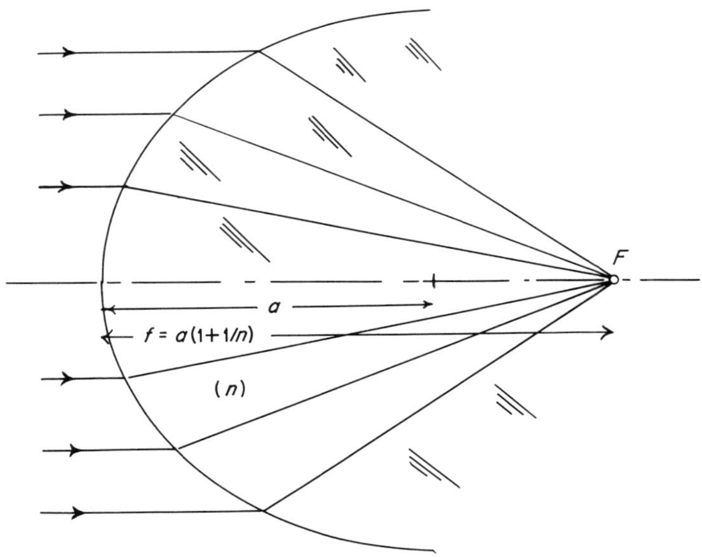

Fig. 3.13 A portion of an ellipsoid of revolution as a single refracting surface free from spherical aberration. The generating ellipse has eccentricity $1/n$ and semimajor axis a. The figure is drawn to scale with $n = 1.81$.

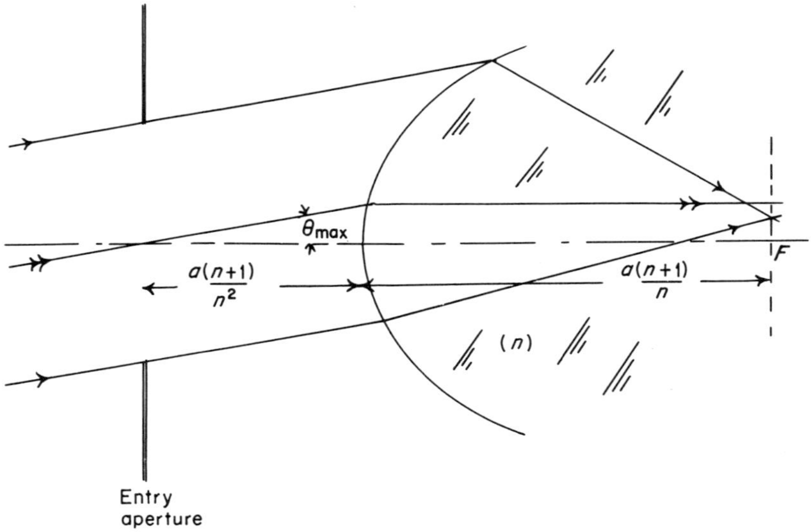

Fig. 3.14 An ellipsoid of revolution as a concentrator. The entry aperture is set at the first focus of the system.

a concentrator we put the entry aperture in front of the ellipsoid, as in Fig. 3.14, so that the principal ray emerges parallel to the axis. We find that the ellipsoid has very strong coma of such a sign that all other rays meet the image plane nearer the axis than the principal ray. Thus all rays strike within this circle. However, this is not an ideal concentrator since, as can be seen from the diagram, the radius of this circle is proportional to tan θ_{max}, whereas, according to Eq. (3.5), all rays should strike within a circle of radius proportional to sin θ_{max}.*

3.4 Media with Continuously Varying Refractive Index

We stated in Section 3.3 that it is thought to be impossible to design an ideal imaging concentrator with a finite number of reflecting or refracting surfaces, even with spherical symmetry, although the example of Fig. 3.10 shows that perfect imagery is possible if one

* In fact, on account of the curvature of the ellipse, the radius is even slightly greater than a value proportional to tan θ_{max}.

conjugate surface is virtual. It has long been known that if we admit continuously varying refractive index, then perfect imagery between surfaces in a spherically symmetric geometry is possible. James Clerk Maxwell (1854) showed that if a medium had the refractive index distribution

$$n = a^2/(b^2 + r^2) \tag{3.9}$$

where a and b are constants and r is a radial coordinate, then any point would be perfectly imaged at another point on the opposite side of the origin. If $a = b = 1$, the distances of conjugate points from the origin are related by

$$rr' = 1 \tag{3.10}$$

and a typical set of imaging rays would be as in Fig. 3.15. This system, known as Maxwell's fisheye lens, is not very useful for our purposes since both object and image have to be immersed in the medium. Luneburg (1964) gave several more examples of media of spherical symmetry with ideal imaging properties. In particular, he found an example, now known in the literature as the Luneburg lens, where the index distribution extends over a finite radius only and where the object conjugate is at infinity. The index distribution is

$$n(r) = \begin{cases} (2 - r^2/a^2)^{\frac{1}{2}}, & r < 1 \\ 1, & r \geq 1 \end{cases} \tag{3.11}$$

This distribution forms a perfect point image with numerical aperture unity, as in Fig. 3.16, and on account of the spherical symmetry it can be shown to form an ideal concentrator of maximum theoretical concentration. We give in Appendix C a more detailed treatment, showing how the ray paths are calculated. There it is shown that the Luneburg lens satisfied Abbe's sine condition [Eq. (3.2)]. Also, it follows from the spherical symmetry that perfect point images are formed from parallel rays coming in all directions. It is then possible to consider the Luneburg lens as having theoretical maximum concentration ratio for any desired collection angle $\theta_{\max}$ up to $\pi/4$, but collecting from a concave spherical source at infinity onto a concave spherical absorber attached to the lens. Apart from the practical problem of making the lens, this is rather an artificial configuration, since until now we have been considering plane entry and exit apertures. Yet the Luneburg lens would have an exit aperture in the form

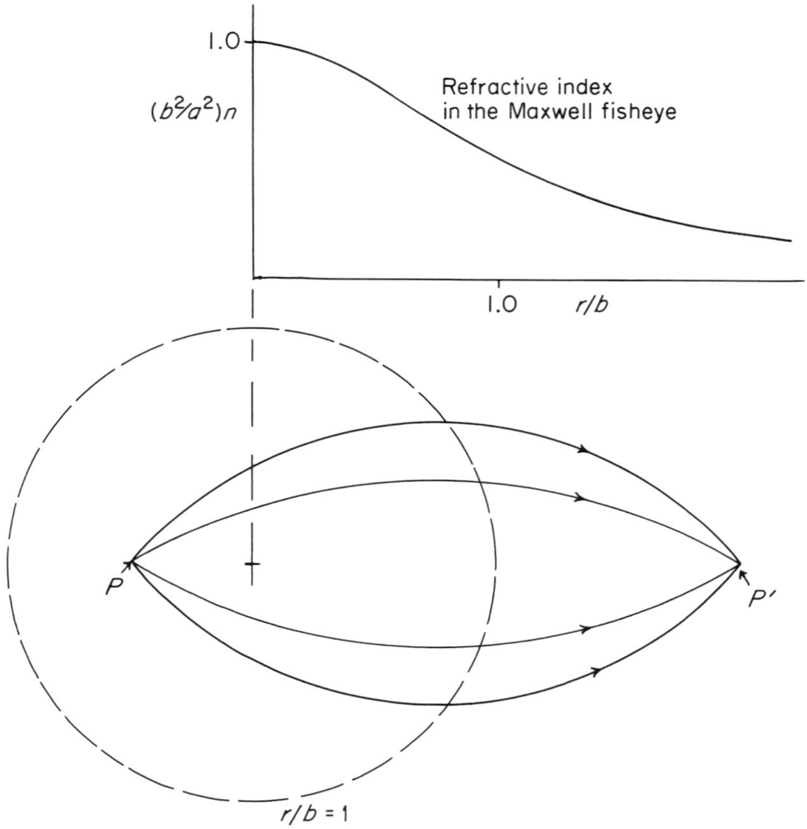

Fig. 3.15 Rays in the Maxwell fisheye. The rays are arcs of circles.

of a spherical cap and an entrance aperture that changes in shape with the angle of the rays. Nevertheless, we show in Appendix C that with reasonable and consistent interpretations of "entrance aperture" and "exit aperture" the Luneburg lens has an optical concentration ratio equal to the theoretical maximum.

3.5 Another System of Spherical Symmetry

The discussion in the last section, in which it was suggested that the ideas of concentration could be extended to nonplane absorbers,

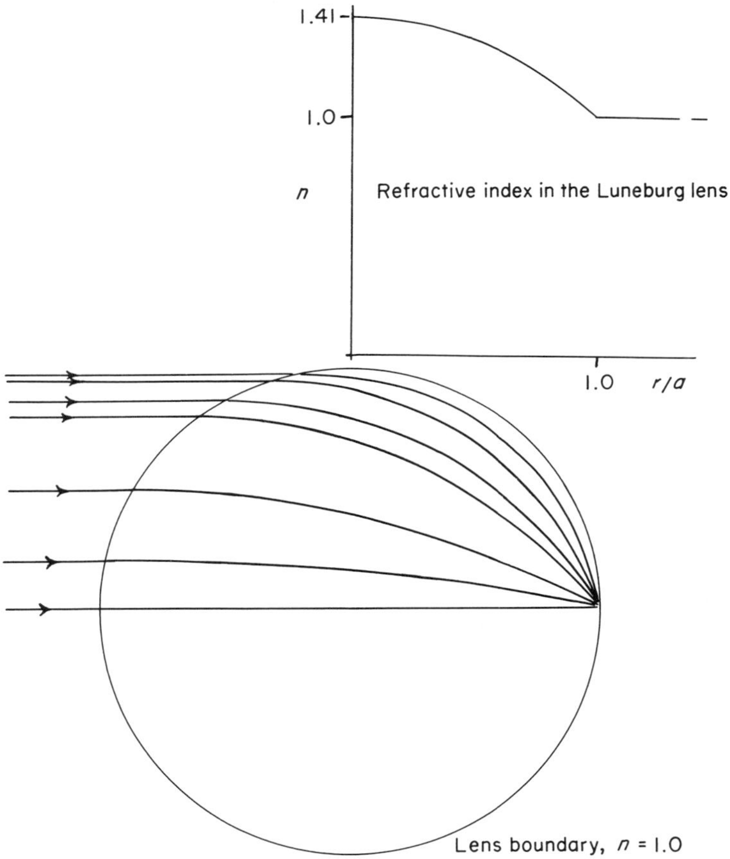

Fig. 3.16 The Luneburg lens.

suggests a way in which the aplanatic imaging system of Fig. 3.10 could be used by itself as a concentrator, as in Fig. 3.17. The surface of radius a/n forms the spherical exit surface, and the internal angle 2α of the cone meeting this face is such that $\sin\alpha = 1/n$. Thus the emerging rays cover a solid angle 2π, as with the Luneburg lens. The entry aperture is now a virtual aperture on the surface of radius *an*. The collecting angle θ_{max} is thus given by $\sin\theta_{max} = 1/n^2$. The concentration ratio from air to air is n^4 for a 3D system, i.e., it is determined only by the refractive index. Similarly, the collecting angle is fixed.

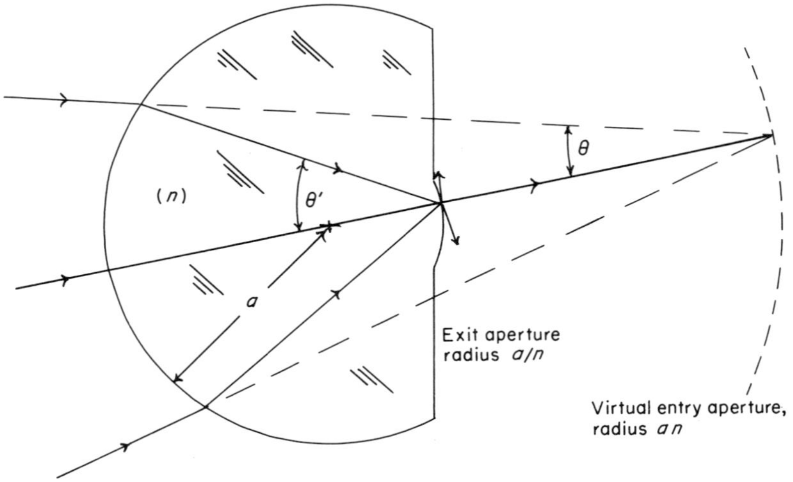

Fig. 3.17 The aplanatic spherical lens as an ideal concentrator; the diagram is to scale for refractive index $n = 2$; the rays shown emerge in air as the extreme rays in a solid angle 2π.

Thus this is not a very flexible system, apart from the fact that it has a virtual collecting aperture. But it does have the theoretical maximum concentration ratio and, if we admit such systems, it is another example of an ideal system.

3.6 Image-Forming Mirror Systems

In this section we examine the performance of mirror systems as concentrators. Concave mirrors have, of course, been used for many years as collectors for solar furnaces and the like. Historical material about such systems is given, e.g., by Krenz (1976). However, little seems to have been published in the way of angle-transmission curves for such systems. Consider first a simple paraboloidal mirror as in Fig. 3.18. As is well known, this mirror focuses rays parallel to the axis exactly to a point focus, or in our terminology it has no spherical aberration. However, the off-axis beams are badly aberrated. Thus in the meridian section (the section of the diagram) it is easily shown by ray tracing that the edge rays at angle θ meet the focal plane farther from the axis than the central ray, so that this cannot be an ideal

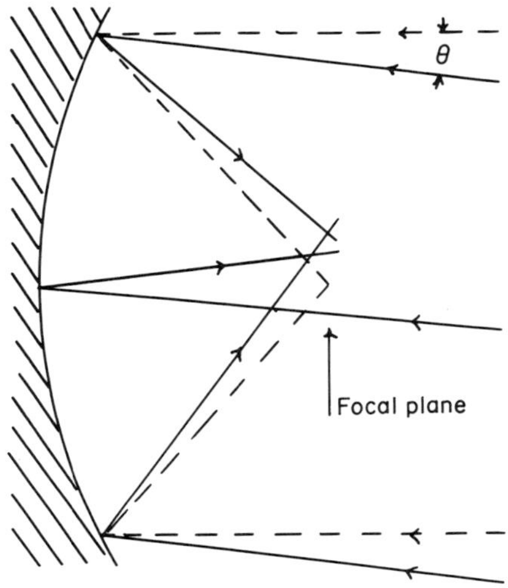

Fig. 3.18 Coma of a paraboloidal mirror. The rays of an axial beam are shown in broken line. The outer rays from the oblique beams at angle θ meet the focal plane farther from the axis than the central ray of this beam.

concentrator even for emergent rays at angles much less than $\pi/2$. An elementary geometrical argument [see, e.g., Harper *et al.* (1976)] shows how big the exit aperture must be to collect all the rays in the meridian section. Referring to Fig. 3.19, we draw a circle passing through the ends of the mirror and the absorber (i.e., exit aperture).

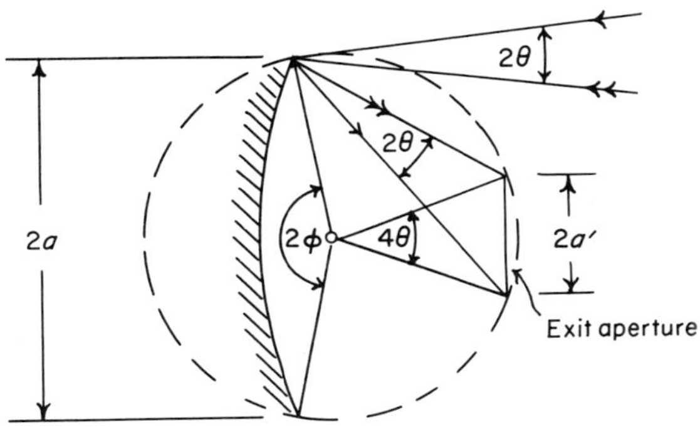

Fig. 3.19 Collecting all the rays from a concave mirror.

Then, by a well-known property of the circle, if the absorber subtends an angle $4\theta_{max}$ at the center of the circle, it subtends $2\theta_{max}$ at the ends of the mirror, so that the collecting angle is $2\theta_{max}$. The mirror is not specified to be of any particular shape except that it must reflect all inner rays to the inside of the exit aperture. Then if the mirror subtends 2φ at the center of the circle we find

$$a'/a = \sin 2\theta_{max}/\sin \varphi \tag{3.12}$$

and the minimum value of a' is clearly attained when $\varphi = \pi/2$. At this point the optical concentration ratio is, allowing for the obstruction caused by the absorber,

$$(a/a')^2 - 1 = [1/(4 \sin^2 \theta_{max})](\cos^2 2\theta_{max}/\cos^2 \theta_{max}) \tag{3.13}$$

It can be seen that this is less than 25% of the theoretical maximum concentration ratio and less than 50% of the ideal for the emergent angle used.

If, as is usual, the mirror is paraboloidal, the rays used for this calculation are actually the extreme rays, i.e., the rays outside of the plane of the diagram all fall within the circle of radius a'.

The large loss in concentration at high apertures is basically because the single concave mirror used in this way has large coma, i.e., it does not satisfy Abbe's sine condition [Eq. (3.2)]. The large amount of coma introduced into the image spreads the necessary size of the exit aperture and so lowers the concentration below the ideal value.

There are image-forming systems that satisfy the Abbe sine condition and have large relative aperture. The prototype of these is the Schmidt camera, which has an aspheric plate and a spherical concave mirror, as in Fig. 3.20. The aspheric plate is at the center of curvature of the mirror, and thus the mirror must be larger than the collecting aperture. Such a system would have the ideal concentration ratio for a restricted exit angle apart from the central obstruction, but there would be practical difficulties in achieving the theoretical maximum. In any case a system of this complexity is clearly not to be considered seriously for solar energy work.

3.7 Conclusions on Image-Forming Concentrators

It must be quite clear by now that, whatever the theoretical possibilities, practical concentrators based on image-forming designs

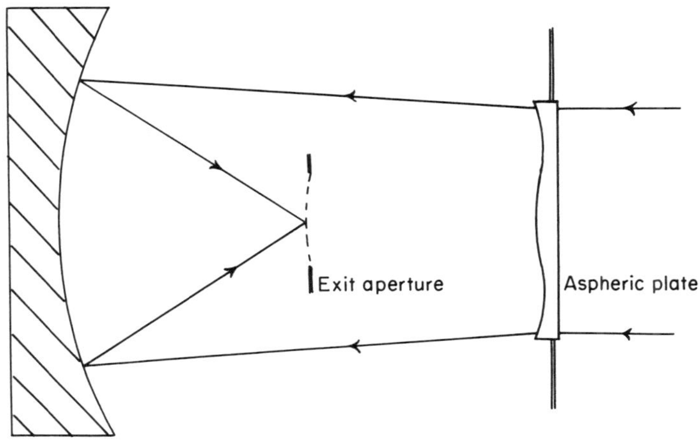

Fig. 3.20 The Schmidt camera. This optical system has no spherical aberration or coma, so that, in principle, it could be a good concentrator for small collecting angles. However, there are serious practical objections, such as cost and the central obstruction of the aperture.

fall a long way short of the ideal. Our graphs of angular transmission will indicate this for some of the simpler designs. As to theoretical possibilities, it is certainly possible to have an ideal concentrator of theoretical maximum concentration ratio if we use a spherically symmetric geometry, a continuously varying refractive index, and quite unrealistic material properties (i.e., refractive index between 1 and 2 and no dispersion). This was proved by the example of the Luneburg lens, and Luneburg and others [e.g. Morgan (1958) and Cornbleet (1976)] have shown how designs suitable for perfect imagery for other conjugates can be obtained.

Perfect concentrators cannot be obtained with plane apertures and axial symmetry if we restrict ourselves to a finite number of elements. However, if we permit unrealistic material properties we can approach indefinitely close to the ideal. In particular, we can get to within diffraction-limited imagery. Probably it is impossible to have a perfect concentrator with a continuously varying index with axial symmetry and plane apertures, but this has not yet been proven rigorously.

CHAPTER 4

Nonimaging Concentrators: The Compound Parabolic Concentrator

4.1 Light Cones

A primitive form of nonimaging concentrator, the light cone, has been used for many years [see, e.g., Holter *et al.* (1962)]. Figure 4.1 shows the principle. If the cone has semiangle γ and if u_{i} is the extreme input angle, then the ray indicated will just pass after one reflection if $2\gamma = (\pi/2) - \theta_{\mathrm{i}}$. It is easy to arrive at an expression for the length of the cone for a given entry aperture diameter. Also, it is easy to see that some other rays incident at angle θ_{i}, such as that indicated by the double arrow, will be turned back by the cone. If we use a longer cone with more reflections, we still find some rays at angle θ_{i} being turned back. Clearly, the cone is far from being an ideal concentrator. Williamson (1952) and Witte (1965) attempted some analysis of the cone concentrator but both restricted this treatment to meridian rays. This unfortunately gives a very optimistic estimate of the concentration. Nevertheless, the cone is very simple compared to the image-forming concentrators described in Chapter 3, and its general form suggests a new direction in which to look for better concentrators.

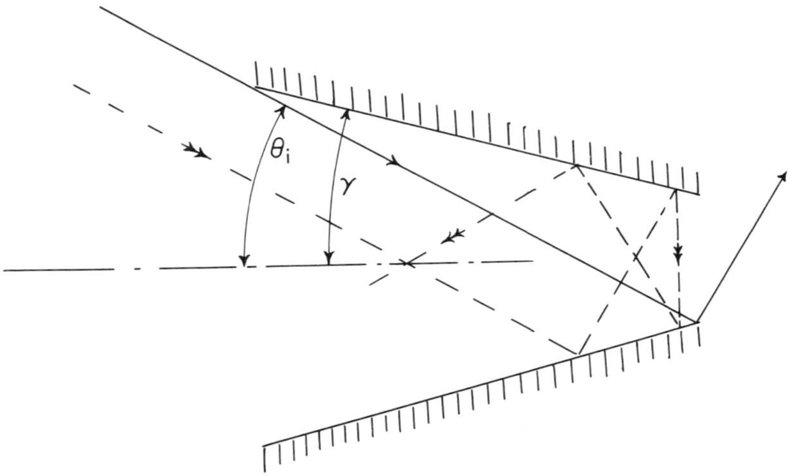

Fig. 4.1 The cone concentrator.

4.2 The Edge-Ray Principle

In Section 3.2 we suggested that an important requirement of an image-forming concentrator is that all rays from the extreme input angle θ_i should form sharp images at the rim of the exit apertures. It seemed reasonable to suppose that rays inside θ_i would then all get through the system and emerge from the exit aperture. Indeed, it is not even necessary that all the extreme rays should form sharp image points, but merely that they should all emerge from the rim of the exit aperture* or, stated more generally, also be the extreme rays at the exit aperture. It does not seem possible to prove that if this "edge-ray principle" is fulfilled the concentrator will be ideal. Also, in the designs of nonimaging concentrators that have been evolved so far there are not enough degrees of freedom to enable us to satisfy the principle for all rays at θ_i. In spite of this, it is found that designs based on the edge-ray principle turn out to have very high concentration ratios, so that it has great heuristic value.

* In image-forming terms this implies astigmatism of the extreme ray pencils, but no spherical aberration or coma.

4.3 The Compound Parabolic Concentrator

If we attempt to improve on the cone concentrator by applying the edge-ray principle, we arrive at the compound parabolic concentrator (CPC), the prototype of a series of nonimaging concentrators that approach very close to being ideal and having the maximum theoretical concentration ratio.

Descriptions of the CPC appeared in the literature in the mid-1960s in widely different contexts. The CPC was described as a collector for light from Čerenkov counters by Hinterberger and Winston (1966a,b). Almost simultaneously, Baranov (1965a) and Baranov and Mel'nikov (1966) described the same principle in 3D geometry and Baranov (1966) suggested 3D CPCs for solar energy collection. Baranov (1965b, 1967) obtained Soviet patents on several CPC configurations. Axially symmetric CPCs were described by Ploke (1967) with generalizations to designs incorporating refracting elements in addition to the light-guiding reflecting wall. Ploke (1969) obtained a German patent for various photometric applications. In other applications to light collection for applications in high-energy physics, Hinterberger and Winston (1966a,b, 1968a,b) noted the limitation to $1/\sin^2 \theta$ of the attainable concentration, but it was not until some time after this that the theory was given explicitly (Winston, 1970). In the latter publication the author derived the generalized étendue (see Appendix A) and showed how the CPC approaches closely to the theoretical maximum concentration.

The CPC in 2D geometry was described by Winston (1974). Further elaborations may be found in Winston and Hinterberger (1975) and Rabl and Winston (1976). Applications of the CPC in 3D form to infrared collection (Harper *et al.*, 1976) and to retinal structure (Winston and Enoch, 1971; Levi–Setti *et al.*, 1975; Baylor and Fettiplace, 1975) have also been described. The general principles of CPC design in 2D geometry are given in a number of U.S. letters patent (Winston, 1975, 1976b, 1977a,b).

Let us now apply the edge-ray principle to improve the cone concentrator. Referring to Fig. 4.2, we require that all rays entering at the extreme collecting angle θ_i shall emerge through the rim point P' of the exit aperture. If we restrict ourselves to rays in the meridian section, the solution is trivial, since it is well known that a parabolic shape with its axis parallel to the direction θ_i and its focus at P' will do this, as in Fig. 4.3. The complete concentrator must have an axis of symmetry

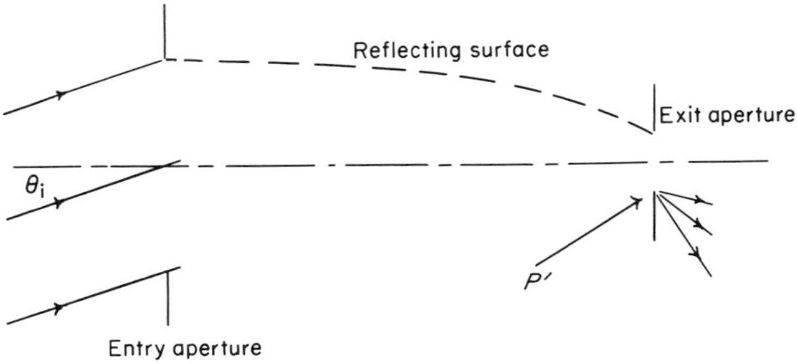

Fig. 4.2 The edge-ray principle.

if it is to be a 3D system, so the reflecting surface is obtained by rotating the parabola about the concentrator axis (*not* about the axis of the parabola).

The symmetry determines the overall length. In the diagram the two rays are the extreme rays of the beam at θ_i, so the length of the concentrator must be such as to just pass both these rays. These considerations determine the shape of the CPC completely in terms of the diameter of the exit aperture $2a'$ and the maximum input angle θ_i. It is a matter of simple coordinate geometry (Appendix D) to show

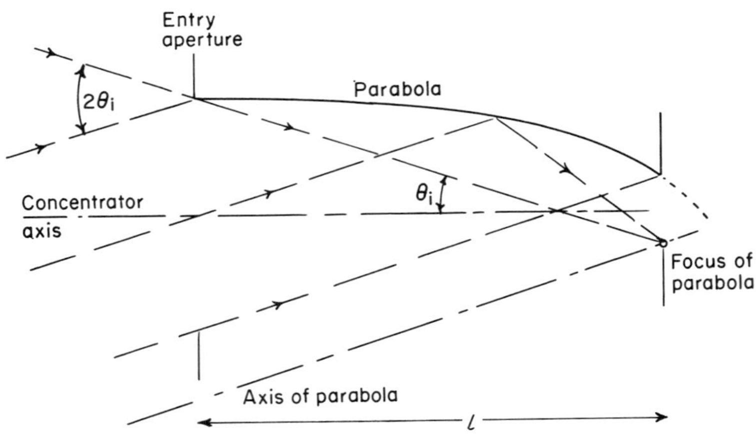

Fig. 4.3 Construction of the CPC profile from the edge-ray principle.

that the focal length of the parabola is

$$f = a'(1+\sin\theta_i) \tag{4.1}$$

the overall length is

$$L = a'(1+\sin\theta_i)\cos\theta_i/\sin^2\theta_i \tag{4.2}$$

and the diameter of the entry aperture is

$$a = a'/\sin\theta_i \tag{4.3}$$

Also, from Eqs. (4.2) and (4.3) or directly from the figure,

$$L = (a+a')\cot\theta_i \tag{4.4}$$

Figure 4.4 shows scale drawings of typical CPCs with a range of collecting angles. It is shown in Appendix D that the concentrator wall has zero slope at the entry aperture, as drawn.

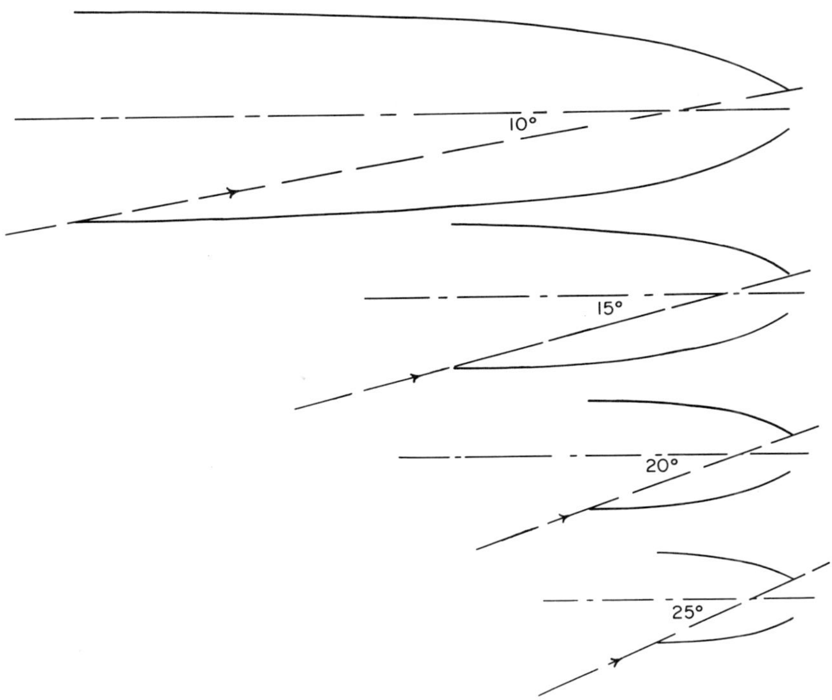

Fig. 4.4 Some CPCs with different collecting angles. The drawings are to scale with the exit apertures all equal in diameter.

The most remarkable result is Eq. (4.3). We see from this that the CPC would have the maximum theoretical concentration ratio (see Section 2.7)

$$a/a' = 1/\sin \theta_i \tag{4.5}$$

provided all the rays inside the collecting angle θ_i actually emerge from the exit aperture. Our use of the edge-ray principle suggests that this ought to be the case, on the analogy with image-forming concentrators, but in fact this is not so. The 3D CPC, like the cone concentrator, has multiple reflections, and these can actually turn back some of the skew rays that enter inside the maximum collecting angle. Nevertheless, the transmission-angle curves for CPCs, as calculated by ray tracing, approach very closely the ideal square shape. Figure 4.5, after Winston (1970), shows a typical transmission-angle curve for a CPC with $\theta_i = 16°$.

It can be seen that the CPC comes very close to being an ideal concentrator. Also, it has the advantages of being a very practical design, easy to make for all wavelengths since it depends on reflection rather than refraction, and of not requiring any extreme material properties. The only disadvantage is that it is very long compared to its diameter, as can be seen from Eq. (4.2). This can be overcome if we incorporate refracting elements into the basic design. In later

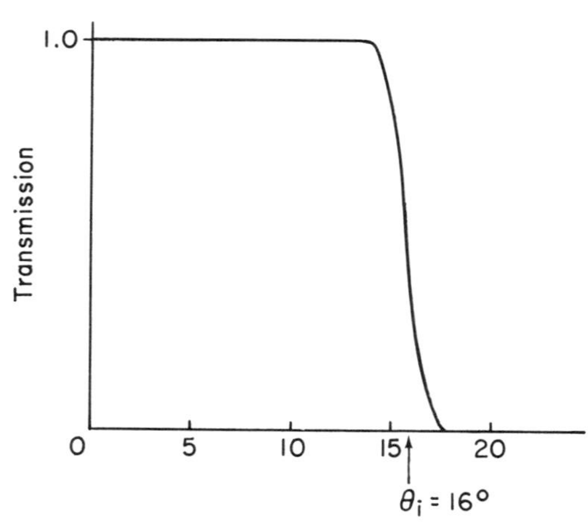

Fig. 4.5 Transmission-angle curve for a CPC with acceptance angle $\theta_i = 16°$. The cut-off occurs over a range of about 1°.

sections of this chapter we shall study the optics of the CPC in detail. We shall elucidate the mechanism by which rays inside the collecting angle are turned back, give transmission-angle curves for several collecting angles, and give quantitative comparisons with some of the other concentrators, imaging and nonimaging, that have been proposed. In later chapters we shall discuss modifications of the basic CPC along various lines, e.g., incorporating transparent refracting materials in the design and even making use of total internal reflection at the walls for all the accepted rays.

We conclude this section by examining the special case of the 2D CPC or troughlike concentrator. This has great practical importance in solar energy applications since, unlike other trough collectors, it does not require diurnal guiding to follow the sun. The surprising result is obtained that the 2D CPC is actually an ideal concentrator of maximum theoretical concentration ratio, i.e., no rays inside the maximum collecting angle are turned back. To show this result, we have to find a way of identifying rays that do get turned back after some number of internal reflections. The following procedure for identifying such rays actually applies not only to CPCs but to all axisymmetric conelike concentrators with internal reflection. It is a way of finding rays on the boundary between sets of rays that are turned back and rays that are transmitted. These extreme rays must just graze the edge of the exit aperture, as in Fig. 4.6, so that if we trace rays in reverse from this point in all directions, as indicated, these rays appear in the entry aperture on the boundary of the required region. Thus we could choose a certain input direction, find the reverse traced rays having this direction, and plot their intersections with the plane of the input aperture. They could be sorted according to the number of reflections involved and the boundaries plotted out. Diagrams of this kind will be given for 3D CPCs in the next chapter.

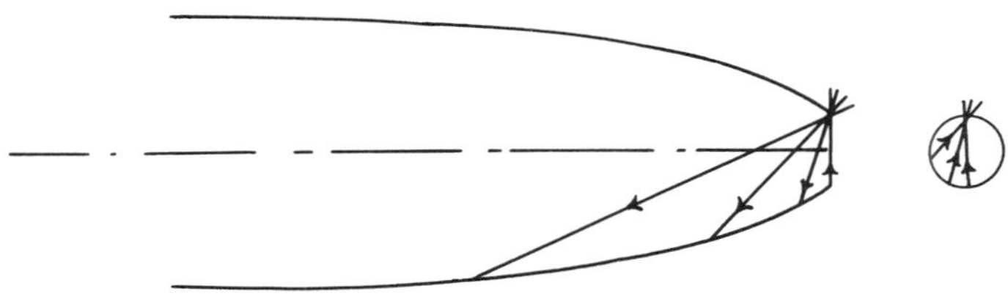

Fig. 4.6 Identifying rays that are just turned back by a conelike concentrator. The rays shown are intended as projections of skew rays, since the meridional rays through the rim correspond exactly to θ_i by construction for a CPC.

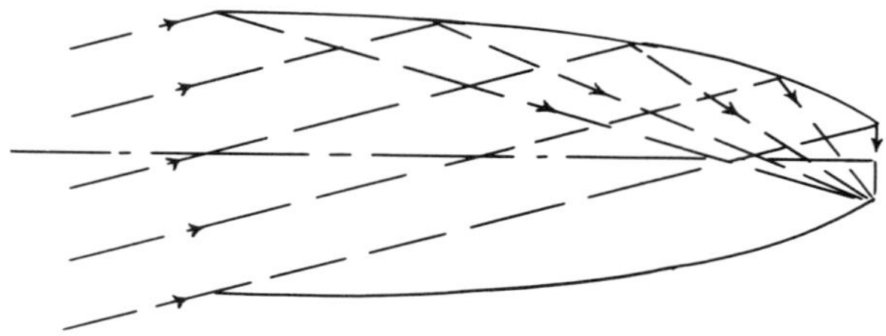

Fig. 4.7 A 2D CPC. The rays drawn represent projections of rays out of the plane of the diagram.

Returning to the 2D CPC, we note first that the ray tracing in any 2D troughlike reflector is simple even for rays not in a plane perpendicular to the length of the trough. This is because the normal of the surface has no component parallel to the length of the trough and thus the law of reflection [Eq. (2.1)] can be applied in two dimensions only. The ray direction cosine in the third dimension is constant. Thus if Fig. 4.7 shows a 2D CPC with the length of the trough perpendicular to the plane of the diagram, all rays can be traced using only their projections on this plane. We can now apply our identification of rays that get turned back. Since, according to the design, all the rays shown appear in the entry aperture at $\theta_{\max}$, there can be no returned rays within this angle. The 2D CPC has maximum theoretical concentration ratio and its transmission-angle graph therefore has the ideal shape, as in Fig. 4.8.*

Since this property is of prime importance, we shall examine the ray paths in more detail to strengthen the verification. Figure 4.9 shows a 2D CPC with a typical ray at the extreme entry angle $\theta_{\max}$. This ray meets the CPC surface at P, say. A neighboring ray at a smaller angle would be represented by the broken line. There are then two possibilities. Either this ray is transmitted as in the diagram, or else it meets the surface again at P_1. In the latter case we apply the same argument but using the extreme ray incident at P_1, and so on. Thus, although some rays have a very large number of reflections, eventually they emerge if they entered inside $\theta_{\max}$. Of course, in the above argument "ray" includes "projection of a ray skew to the diagram."

* Strictly, this applies to 2D CPCs that are indefinitely extended along the length of the trough. In practice, this effect is achieved by closing the ends with plane mirrors perpendicular to the straight generators of the trough. This ensures that *all* rays entering the rectangular entry aperture within the acceptance angle emerge from the exit aperture.

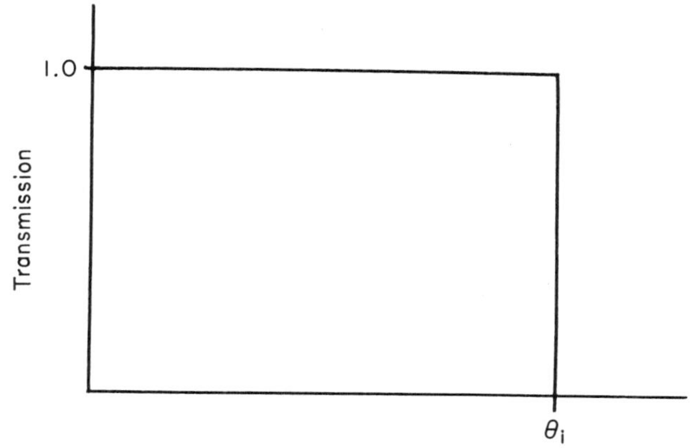

Fig. 4.8 The transmission-angle curve for a 2D CPC.

This result shows up a difference between 2D and 3D CPCs. The 2D CPC has maximum theoretical concentration, in the sense of Section 2.9, but in extending it to 3D we have included more rays (there is now a threefold infinity of rays, allowing for the axial symmetry, whereas in the 2D case we have to consider only a twofold infinity) but we have no more degrees of freedom in the design, since the 3D concentrator is obtained from the 2D profile by rotation about the axis of symmetry. The 3D concentrator has to be a figure of revolution and thus we can do nothing to ensure that rays outside the meridian sections are properly treated. We shall see in Section 4.4.3 that it is precisely the rays in these regions that are turned back by multiple reflections inside the CPC.

This discussion also shows the different causes of nonideal performance of imaging and nonimaging systems. The rays in an image-forming concentrator such as a high-aperture lens all pass through each surface the same number of times (usually once) and the nonideal

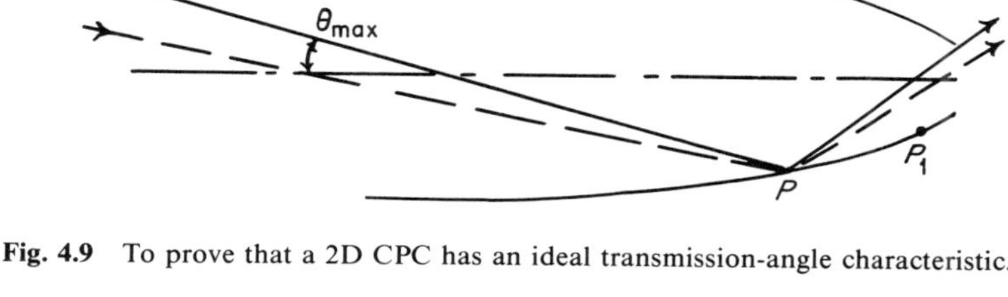

Fig. 4.9 To prove that a 2D CPC has an ideal transmission-angle characteristic.

performance is caused by geometrical aberrations in the classical sense. In a CPC, on the other hand, different rays have different numbers of reflections before they emerge (or not) at the exit aperture. It is the effect of the reflections in turning back the rays that produces nonideal performance. Thus there is an essential difference between a lens with large aberrations and a CPC or other nonimaging concentrator. A CPC is a system of rotational symmetry and it would be possible to consider all rays having just, say, three reflections and discuss the aberrations (no doubt very large) of the image formation by these rays. But there seems no sense in which rays with different numbers of reflections could be said to form an image. It is for this reason that we continue to draw the distinction between image-forming and nonimaging concentrators.

4.4 Properties of the Compound Parabolic Concentrator

In this section we shall give in some detail the properties of the basic CPC of which the design was developed in the last section. We shall show how ray tracing can be done, give the results of ray tracing in the form of transmission-angle curves, state certain general properties of these curves, and show the patterns of rays in the entry aperture that get turned back. This detailed examination will help in elucidating the mode of action of CPCs and their derivatives, to be described in later chapters.

4.4.1 *The Equation of the CPC*

By rotation of axes and translation of origin we can write down the equation of the meridian section of a CPC. In terms of the diameter $2a'$ of the exit aperture and the acceptance angle $\theta_{\max}$ this equation is

$$\begin{aligned}(r\cos\theta_{\max}+z\sin\theta_{\max})^2&\\ +2a'(1+\sin\theta_{\max})^2 r-2a'\cos\theta_{\max}(2+\sin\theta_{\max})z&\qquad(4.6)\\ -a'^2(1+\sin\theta_{\max})(3+\sin\theta_{\max})=0&\end{aligned}$$

where the coordinates are as in Fig. 4.10. Recalling that the CPC is a surface of revolution about the z axis, we see that in three dimensions, with $r^2=x^2+y^2$, Eq. (4.6) represents a fourth-degree surface.

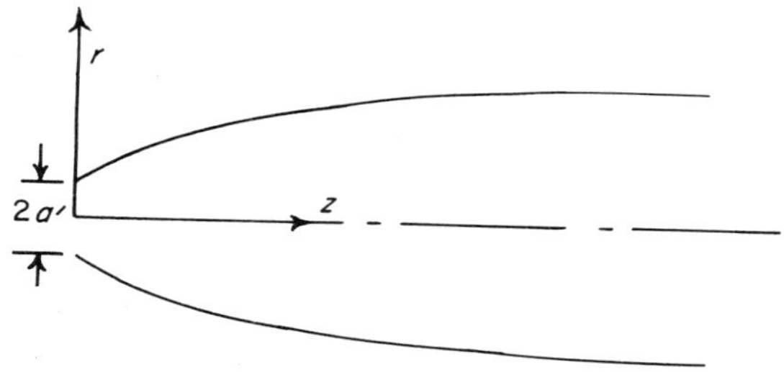

Fig. 4.10 The coordinate system for the r–z equation for the CPC.

A more compact parametric form can be found by making use of the polar equation of the parabola. Figure 4.11 shows how the angle φ is defined. In terms of this angle and the same coordinates (r, z), the meridian section is given by

$$r = \frac{2f\sin(\varphi - \theta_{max})}{1 - \cos\varphi} - a', \qquad z = \frac{2f\cos(\varphi - \theta_{max})}{1 - \cos\varphi} \tag{4.7}$$

$[f = a'(1 + \sin\theta_{max})]$.

If we introduce an azimuthal angle ψ, we obtain the complete parametric equations of the surface:

$$\begin{aligned} x &= \frac{2f\sin\psi\sin(\varphi - \theta_{max})}{1 - \cos\varphi} - a'\sin\psi \\ y &= \frac{2f\cos\psi\sin(\varphi - \theta_{max})}{1 - \cos\varphi} - a'\cos\psi \\ z &= \frac{2f\cos(\varphi - \theta_{max})}{1 - \cos\varphi} \end{aligned} \tag{4.8}$$

The derivations of these equations are sketched in Appendix D.

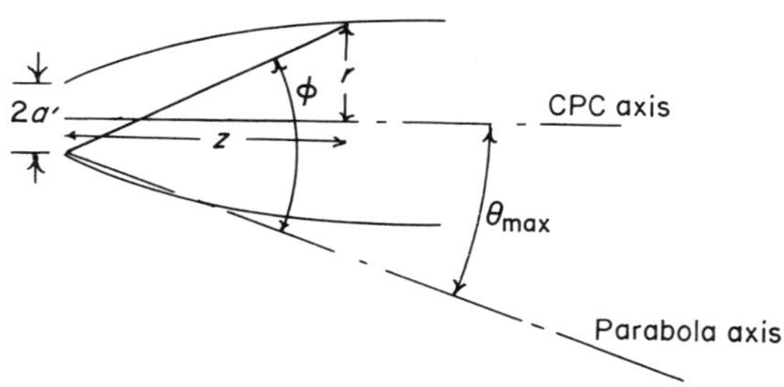

Fig. 4.11 The angle φ used in the parametric equations of the CPC.

4.4.2 *The Normal to the Surface*

We need the direction cosines of the normal to the surface of the CPC for ray-tracing purposes. There are well-known formulas of differential geometry that give these. If the explicit substitution $r = (x^2 + y^2)^{\frac{1}{2}}$ is made in Eq. (4.6) and the result is written in the form

$$F(x, y, z) = 0 \tag{4.9}$$

the direction cosines are given by

$$\mathbf{n} = (F_x, F_y, F_z)/(F_x^2 + F_y^2 + F_z^2)^{\frac{1}{2}} \tag{4.10}$$

The formulas for the normal are slightly more complicated for the parametric form. We first define the two vectors

$$\mathbf{a} = (\partial x/\partial\varphi, \partial y/\partial\varphi, \partial z/\partial\varphi), \qquad \mathbf{b} = (\partial x/\partial\psi, \partial y/\partial\psi, \partial z/\partial\psi) \tag{4.11}$$

Then the normal is given by

$$\mathbf{n} = \mathbf{a} \times \mathbf{b}/\{|\mathbf{a}|^2|\mathbf{b}|^2 - |\mathbf{a} \cdot \mathbf{b}|^2\}^{\frac{1}{2}} \tag{4.12}$$

These results are given in elementary texts such as that by Weatherburn (1931).

Although the formulas for the normal are somewhat opaque, it can be seen from the construction for the CPC profile in Fig. 4.3 that at the entry end the normal is perpendicular to the CPC axis, i.e., the wall is tangent to a cylinder.

4.4.3 *Transmission-Angle Curves for CPCs*

In order to compute the transmission properties of a CPC, the entry aperture was divided into a grid with spacing equal to 1/100 of the diameter of the aperture, and rays were traced at a chosen collecting angle θ at each grid point. The proportion of these rays that were transmitted by the CPC gave the transmission $T(\theta, \theta_{\max})$ for the CPC with maximum collecting angle $\theta_{\max}$. $T(\theta, \theta_{\max})$ was then plotted against θ to give the transmission-angle curve. Some of these curves are given in Fig. 4.12. They all approach very closely the ideal rectangular cut-off that a concentrator with maximum theoretical concentration ratio should have. The transition from $T = 0.9$ to $T = 0.1$

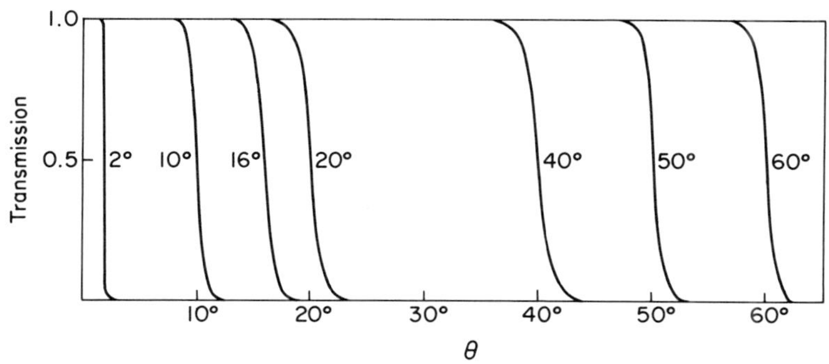

Fig. 4.12 Transmission-angle curves for 3D CPCs with θ_{max} from 2° to 60°.

takes place in $\Delta\theta$ less than 3° in all cases. Approximate values are

θ_{max}	2°	10°	16°	20°	40°	60°
$\Delta\theta$	0.4°	1.5°	2°	2.5°	2.7°	2.0°

We may also be interested in the total flux transmitted inside the design collecting angle θ_{max}. This is clearly proportional to

$$\int_0^{\theta_{max}} T(\theta, \theta_{max}) \sin 2\theta \, d\theta \tag{4.13}$$

and if we divide by $\int_0^{\theta_{max}} \sin 2\theta \, d\theta$, we obtain the fraction transmitted of the flux incident inside a cone of semiangle θ_{max}. The result of such a calculation is shown in Fig. 4.13. This gives the proportion by which the CPC fails to have the theoretical maximum concentration ratio. For example, the 10° CPC should have the theoretical concentration ratio $\operatorname{cosec}^2 10° = 33.2$, but from the graph it will actually have 32.1. The loss is, of course, because some of the skew rays have been turned back by multiple reflections inside the CPC.

It is of some considerable theoretical interest to see how these failures occur. By tracing rays at a fixed angle of incidence, regions could be plotted in the entry aperture showing what happened to rays in each region. Thus Fig. 4.14 shows these plots for a CPC with $\theta_{max} = 10°$ for rays at 8°, 9°, 9.5°, 10°, 10.5°, 11°, 11.5°. Rays incident in regions labeled 0, 1, 2, . . . are transmitted by the CPC after zero, one, two, . . . reflections; $F2$, $F3$, . . . indicate that rays incident in those regions begin to turn back after two, three, . . . reflections. Rays in the

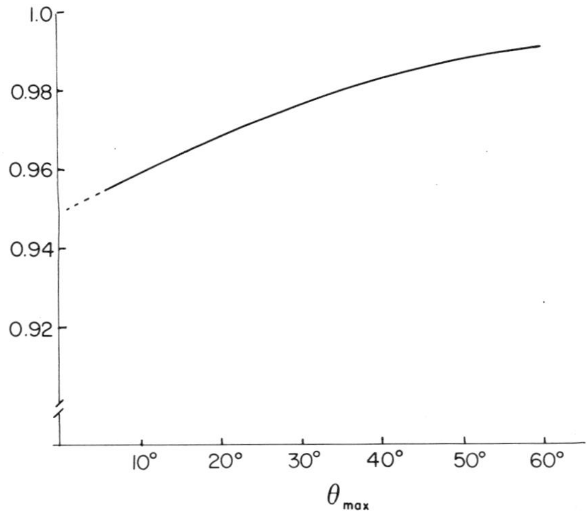

Fig. 4.13 Total transmission within $\theta_{\max}$ for 3D CPCs.

blank regions will still be traveling toward the exit aperture after five reflections. The calculations were abandoned here to save computer time. In the computation of $T(\theta, \theta_{\max})$ these rays were omitted. For θ less than $\theta_{\max}$ it is most likely that all these rays are transmitted, as we shall show below, but for θ greater than the $\theta_{\max}$ this is probably not so. Thus the transitions in the curves of Fig. 4.12 are probably slightly sharper than shown.

The boundaries between regions in the diagrams of Fig. 4.14 are, of course, distorted images of the exit aperture seen after various numbers of reflections. It can be seen that the failure regions, i.e., regions in which rays are turned back, appear as a splitting between these boundary regions. For example, the regions for failure after two and three reflections for 9° appear in the diagram as a split between the regions for transmission after one and two reflections. This confirms the principle stated in Section 4.3 that rays that meet the rim of the exit aperture are at the boundaries of failure regions. Naturally enough, each split between regions for transmission after n and $n+1$ reflections produces two failure regions, for failure after $n+1$ and $n+2$ reflections.

We can delineate these regions in another way, by tracing rays in reverse from the exit aperture. Thus in Fig. 4.15 we can trace rays in

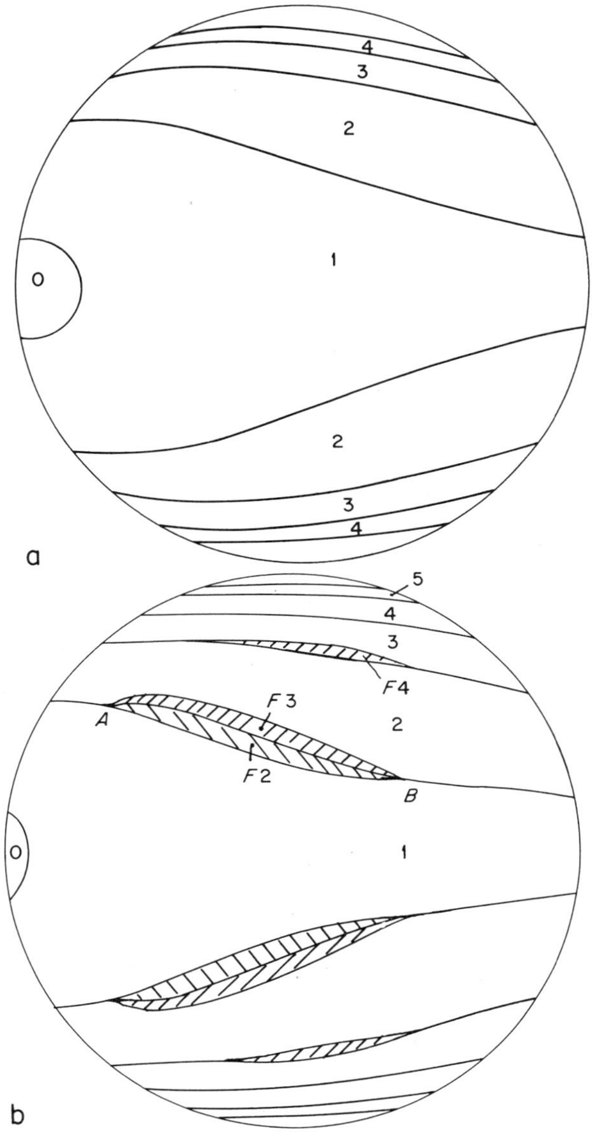

Fig. 4.14 Patterns of accepted and rejected rays at the entry face of a 10° CPC. The entry aperture is seen from above with incident rays sloping downward to the right. Rays entering areas labeled n are transmitted after n reflections; those entering hatched areas labeled Fm are turned back after m reflections. The ray trace was not carried to completion in the unlabeled areas. (a) 8°, $\theta_{max} = 10°$; (b) 9°, $\theta_{max} = 10°$; (c) 9.5°, $\theta_{max} = 10°$; (d) 10°, $\theta_{max} = 10°$; (e) 10.5°, $\theta_{max} = 10°$; (f) 11°, $\theta_{max} = 10°$; (g) 11.5°, $\theta_{max} = 10°$. (*Figure continues*)

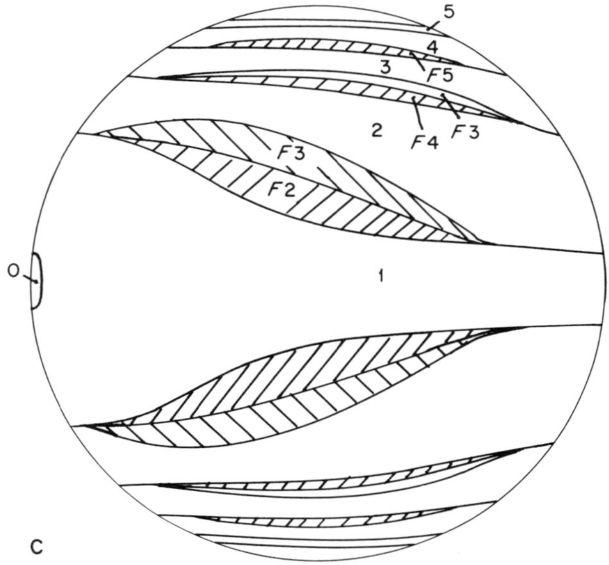

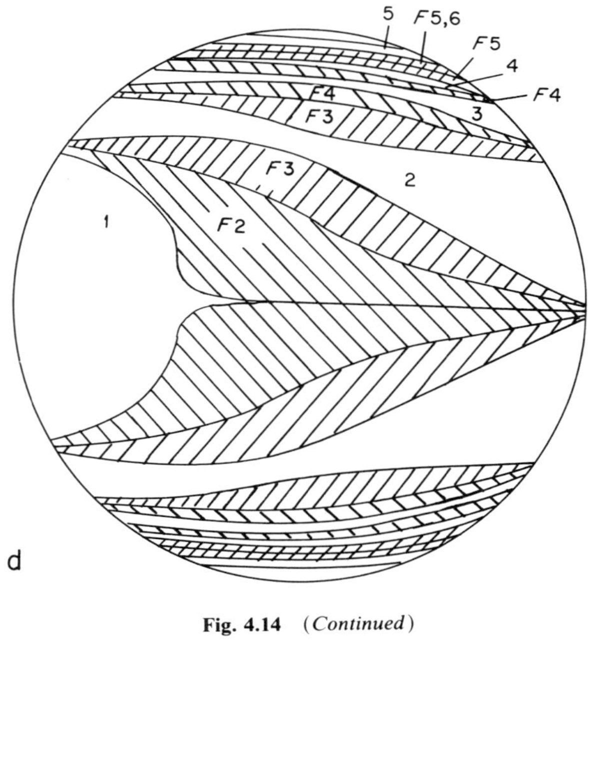

Fig. 4.14 (*Continued*)

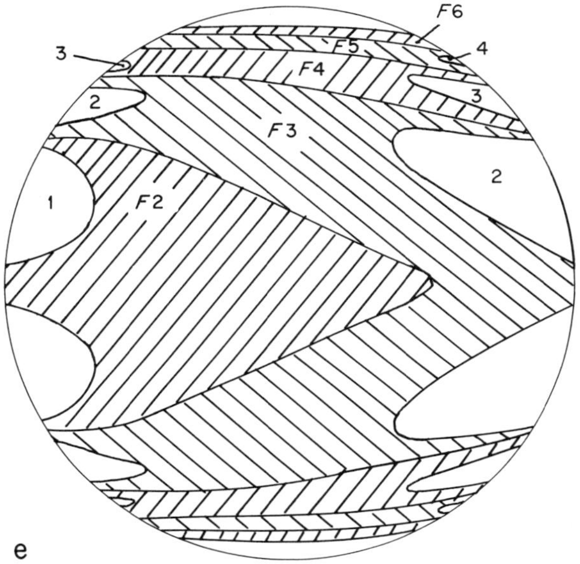

e

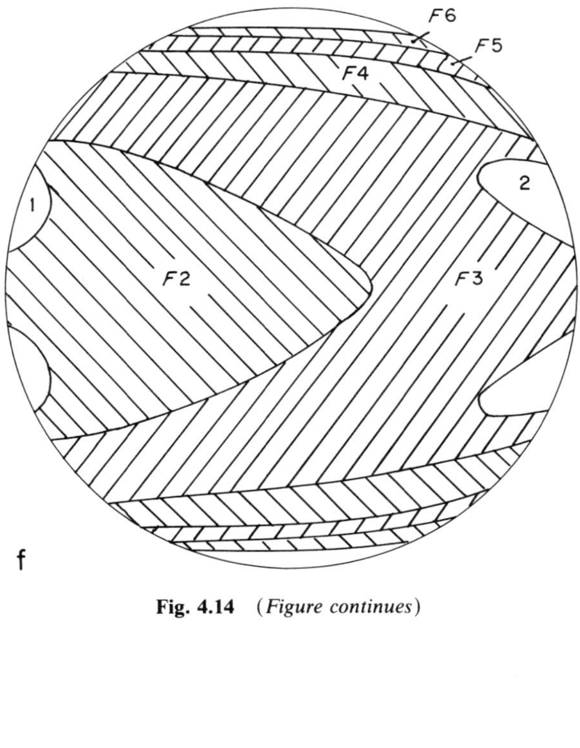

f

Fig. 4.14 (*Figure continues*)

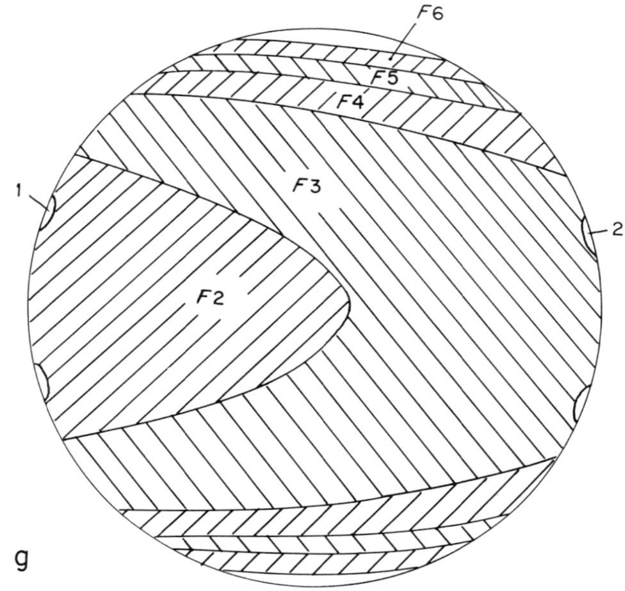

Fig. 4.14 (*Continued*)

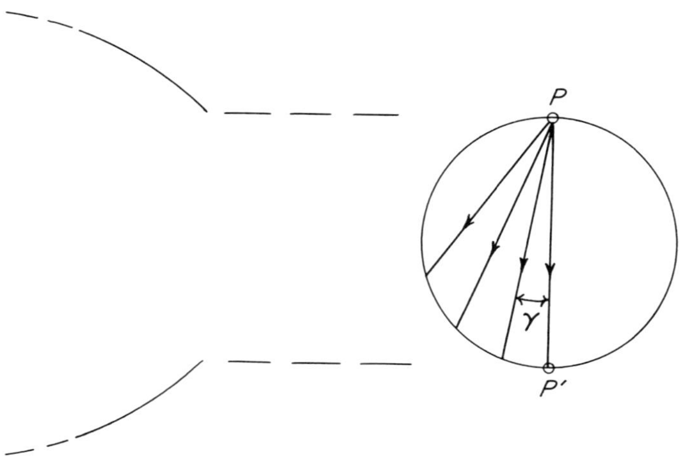

Fig. 4.15 Rays at the exit aperture used to determine failure regions.

the plane of the exit aperture from a point P at angles γ to the diameter P'. Each ray will eventually emerge from the entry face at a certain angle $\theta(\gamma)$ to the axis and after n reflections. The point in the entry aperture from which this ray emerges is then the point in diagrams, such as those of Fig. 4.14, at which the split between rays transmitted after $n-1$ and n reflections begins. For example, to find the points A and B in the 9° diagram of Fig. 4.14, we look for an angle γ that yields $\theta(\gamma)=9°$ and find the coordinates of the ray emerging from the entry face after two reflections. There will, of course, be two such values of γ, corresponding to the two points A and B. This was verified by ray tracing.

Returning to the blank regions in Fig. 4.14, the rays entering at these regions are almost tangential to the surface of the CPC. Thus they will follow a spiral path down the CPC with many reflections, as indicated in Fig. 4.16. We can use the skew invariant h explained in Section 2.8 to show that such rays must be transmitted if the incident angle is less than θ_{max}. For if we use the reversed rays and take a ray with $\gamma=\pi/2$ as in Fig. 4.15, this ray has $h=a'$. When it has spiraled back to the entry end (with an infinite number of reflections!) it must have the same $h=a'=a' \sin \theta_{max}$, i.e., it emerges tangent to the CPC

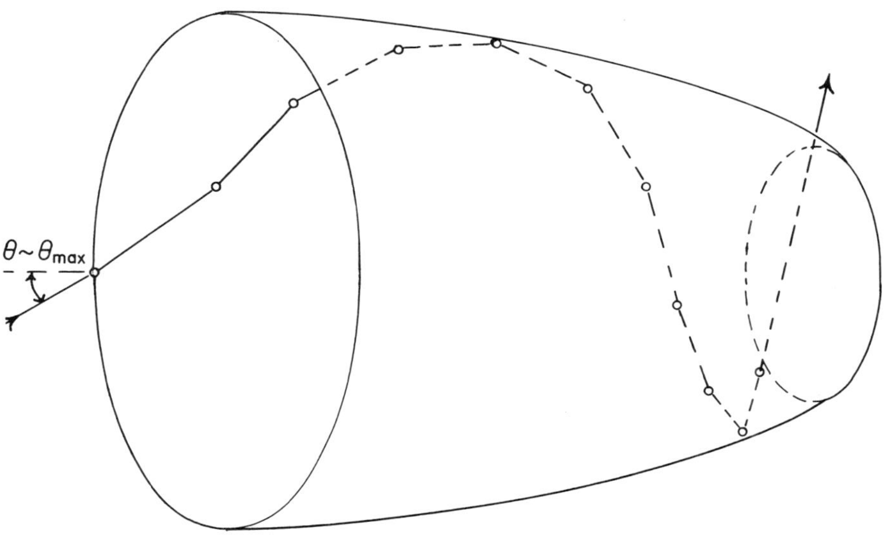

Fig. 4.16 Path of a ray striking the surface of a CPC almost tangentially.

surface at the maximum collecting angle. Any other ray in the blank regions closer to the axis or with smaller θ has a smaller skew invariant and is therefore transmitted.

The above argument holds for regions very close to the rim of the CPC. The reverse ray-tracing procedure shows that for angles below θ_{max} the failures begin well away from the blank region—in fact, at approximately half the radius at the entry aperture. Thus we are justified in including the blank regions in the count of rays passed for $\theta < \theta_{max}$, as suggested above. This argument also shows that the transmission-angle curves (Fig. 4.12) are precisely horizontal out to a few degrees below θ_{max}.

A converse argument shows, on the other hand, that rays incident in this region at angle above θ_{max} will not be transmitted. There seems to be no general argument to show whether or not the transmission goes precisely to zero at angles sufficiently greater than θ_{max}, but the ray-tracing results suggest very strongly that it does.

4.5 Cones and Paraboloids as Concentrators

Cones are much easier to manufacture than CPCs. Paraboloids of revolution (which of course CPCs are not) seem a more natural choice to conventional optical physicists as concentrators. We therefore give in this section some quantitative comparisons. It will appear from these that the CPC has very much greater efficiency as a concentrator than either of these other shapes.

In order to make a meaningful comparison, the concentration ratio as defined by the ratio of the entrance and exit aperture areas was made the same as for the CPC with $\theta_{max} = 10°$, i.e., a ratio of 5.76 to 1 in diameter. The length of the cone was chosen so that the ray at θ_{max} was just cut off, as in Fig. 4.17. For the paraboloid, the exit aperture diameter and the concentration ratio completely determine the shape, as in Fig. 4.18.

Figures 4.19 and 4.20 show the transmission-angle curves for cones and paraboloids, respectively. It is obvious that the characteristics of both these systems as concentrators are much worse than ideal. For example, the total transmission inside θ_{max} for the paraboloids, according to Eq. (4.13), is about 0.60 for all the angles shown. The cones clearly have definitely better characteristics than the paraboloids, with

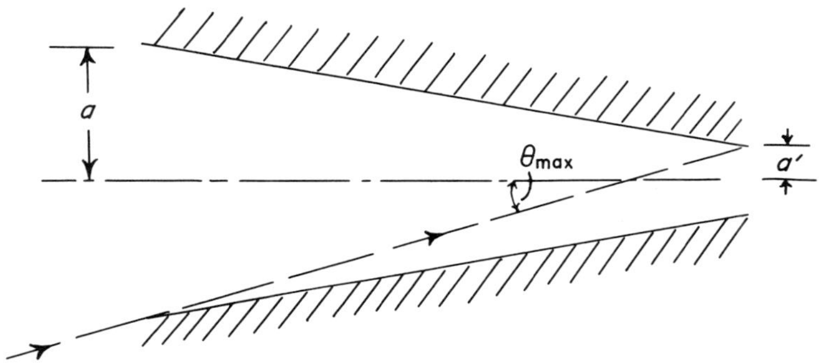

Fig. 4.17 A cone concentrator, showing dimensions used to compare with a CPC.

a total transmission inside θ_{max} of order 80%. This is perhaps a verification of our view that nonimaging systems can have better concentration than image-forming systems, since the paraboloid of revolution is an image-forming system, albeit with very large aberrations when used in the present way.*

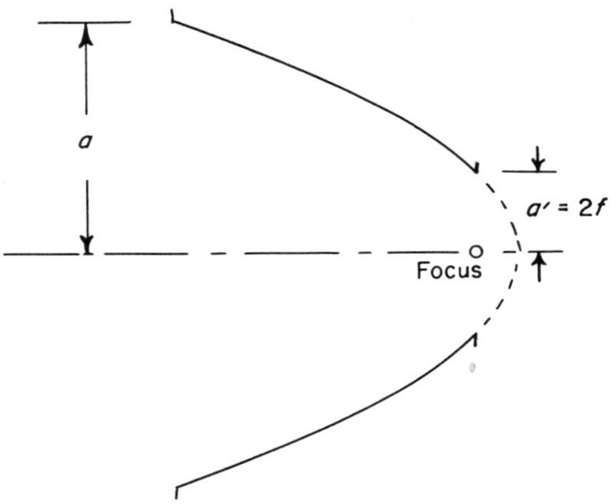

Fig. 4.18 A paraboloid of revolution as a concentrator.

* The section of a paraboloid of revolution in front of the focus is used for X-ray imaging, since most materials are good reflectors for X rays at grazing incidence.

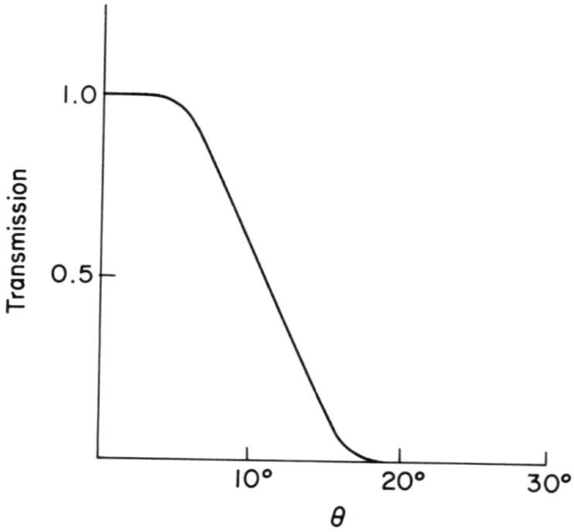

Fig. 4.19 Transmission-angle curve for a cone; $\theta_{max} = 10°$.

Rabl (1976b) considered V-troughs, i.e., 2D cones, and used a well-known construction (e.g., Williamson, 1952) shown in Fig. 4.21 to estimate the angular width of the transition region in the transmission-angle curve. He showed that its width was equal to the angle 2φ of the V-trough and the center of the transition same at $\delta + \varphi$, where δ is the largest angle of an incident pencil for which all rays are transmitted. If we assume the same holds for a 3D cone, it suggests

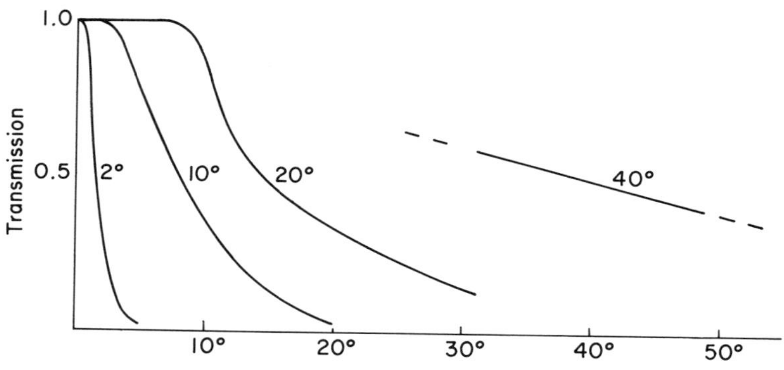

Fig. 4.20 Transmission-angle curves for paraboloidal mirrors. The graphs are labeled with angles θ_{max} given by $\sin \theta_{max} = a'/a$ in Fig. 4.18.

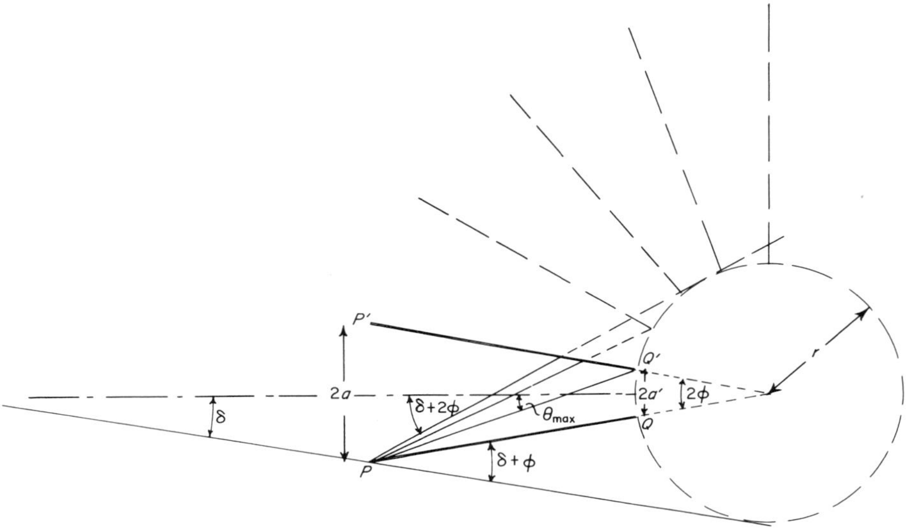

Fig. 4.21 The pitch-circle construction for reflecting cones and V-troughs. A straight line through the entry aperture of the V-trough emerges from the exit aperture if it cuts the pitch circle. Otherwise, it is turned back. The construction is only valid for meridian rays of a cone.

that the transition in the transmission-angle curve becomes sharper as the cone angle decreases, i.e., smaller angle cones are more nearly ideal concentrators. This accords with Garwin's result (see Appendix B), which may be said to imply that a very long cone with a very small angle is a nearly ideal concentrator.

A different way to look at the performance of the cone is to note that for small φ the concentration ratio a/a' is approximately $1/\sin(\delta + \varphi)$. Thus as the cone length increases while a/a' is held fixed, φ/δ tends to zero, and from the diagram the transition region in the transmission-angle diagram becomes sharper. Nevertheless, there is always a finite transition even for V-troughs and more so for cones , so that the comparison with the CPC always shows that the cone is much less efficient and departs further from the ideal than the CPC.

Finally, Fig. 4.22 shows the pattern of rays accepted and rejected by a 10° cone as seen at the entry aperture. This may be compared with Fig. 4.14d, which shows the pattern for a 10° CPC.

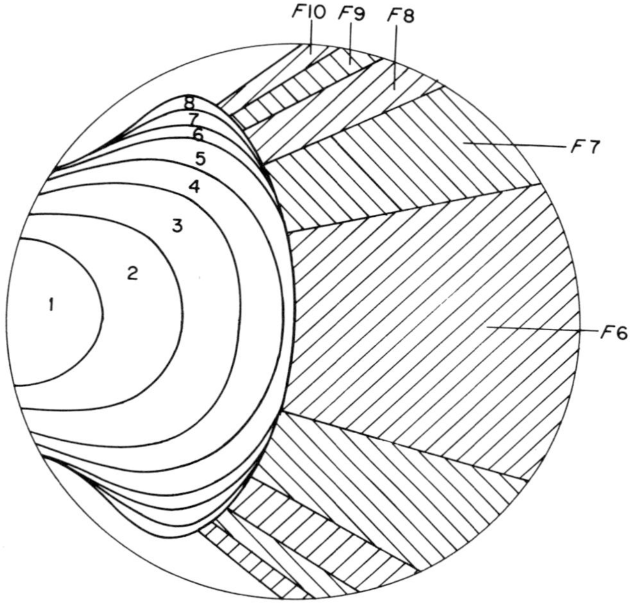

Fig. 4.22 Patterns of accepted and rejected rays at the entry aperture of a 10° cone for rays at 10°. The notation is as for Fig. 4.14, and this figure may be compared directly with Fig. 4.14d.

CHAPTER 5

Developments and Modifications of the Basic Compound Parabolic Concentrator

5.1 Introduction

There are several possible ways in which the basic CPC as described in Chapter 4 could be varied for specific purposes. Some of these have been hinted at already in this book, e.g., a solid dielectric CPC using total internal reflection. Others spring to mind fairly readily for specific purposes, e.g., collecting from a source at a finite distance rather than at infinity. In this chapter we shall describe these developments and discuss their properties.

5.2 The Dielectric-Filled CPC with Total Internal Reflection

Both 2D and 3D CPCs filled with dielectric and using total internal reflection were described by Winston (1976a). If we consider either the 2D case or meridian rays in the 3D case, it can be seen that the minimum angle of incidence for rays inside the design collecting angle occurs at the rim of the exit aperture, as in Fig. 5.1. If the dielectric has refractive index n, the CPC is, of course, designed with an acceptance angle θ' inside the dielectric, according to the law of refraction.

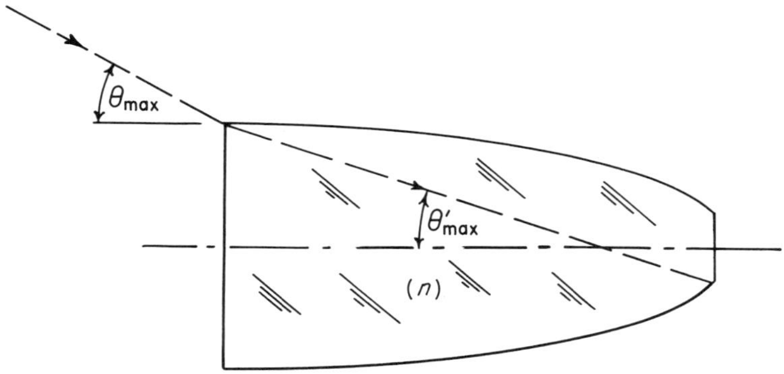

Fig. 5.1 A dielectric-filled compound parabolic concentrator. The figure is drawn for an entry angle of 18° and a refractive index of 1.5. The concentration ratio is thus 10.2 for a 3D concentrator.

It is then easy to show that the condition for total internal reflection to occur at all points is

$$\sin\theta'_i \le 1-(2/n^2) \qquad \text{or} \qquad \sin\theta_i \le n-(2/n) \tag{5.1}$$

Since the sine function can only take values between 0 and 1, it can be seen that the useful values of n are greater than $\sqrt{2}$. This is in good agreement with the range of useful optical materials in the visible and infrared regions. In a 3D CPC, rays outside the meridian plane have a larger angle of incidence than meridian rays at the same inclination to the axis, so that Eq. (5.1) covers all CPCs. The expressions in Eq. (5.1) are plotted in Fig. 5.2. For most purposes it is unlikely that collecting angles exceeding 40° would be needed, so that the range of useful n coincides very well. For trough collectors there is always total internal reflection at the perpendicular end walls, since there is the same angle of incidence here as for the ray at θ_i at the entry aperture on the curved surface. In fact, it could be shown that for $n > \sqrt{2}$ it is impossible for a ray to get into a trough CPC and not be totally internally reflected at the end face.

The angular acceptance of the dielectric-filled 2D CPC for non-meridional rays is actually larger than a naive analogy with the $n=1$ case would indicate (Scharlack, 1977). To see this, it is convenient to represent the angular acceptance by the direction cosine variables introduced earlier. Let x be transverse to the trough, let y lie along

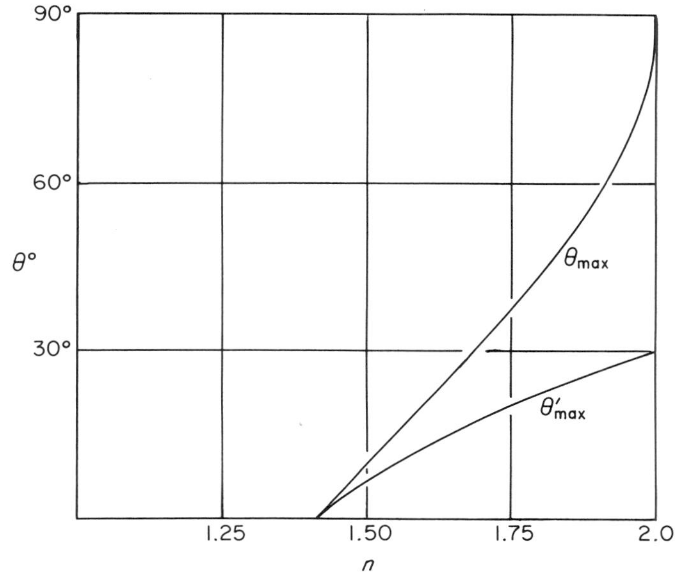

Fig. 5.2 The maximum collecting angles for a dielectric-filled CPC with total internal reflection, as functions of the refractive index.

the trough, and let L and M be the corresponding direction cosines. We recall that the ordinary $(n=1)$ 2D CPC accepts all rays whose projected angles in the xz plane are $\leq \theta_i$, the design cut-off angle. This condition is represented by an ellipse in the L, M plane with semidiameters $\sin\theta_i$ and 1, respectively (shown in Fig. 5.3a). Therefore, the acceptance figure just inside the dielectric is such an ellipse with semidiameters $\sin\theta_i$ and 1. In terms of direction cosines, Snell's law is simply

$$L = nL', \qquad M = nM' \tag{5.2}$$

so one might expect the acceptance ellipse in the L, M plane to be scaled up by n. However, physical values of L and M lie inside the unit circle

$$L^2 + M^2 < 1 \tag{5.3}$$

It follows that the accepted rays lie inside the intersection of the scaled-up ellipse and the unit circle. This region, as seen in Fig. 5.3b,

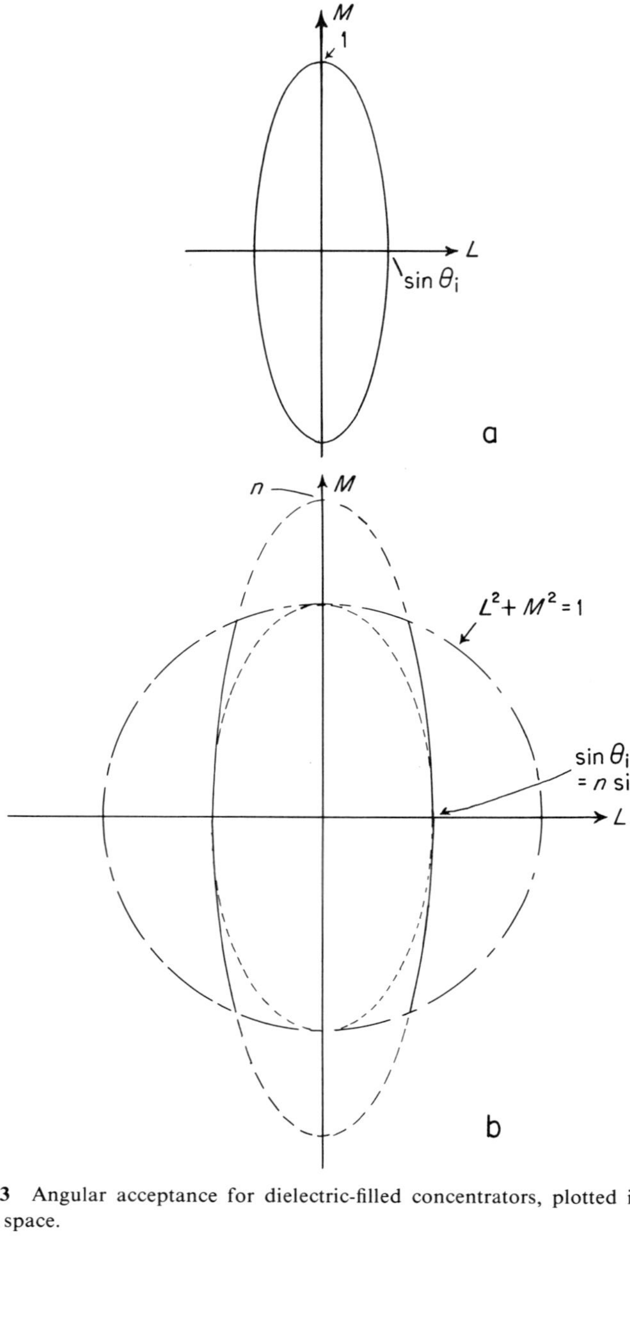

Fig. 5.3 Angular acceptance for dielectric-filled concentrators, plotted in direction cosine space.

is larger than an ellipse with semidiameters $\sin\theta_i = n\sin\theta_i'$ and 1, respectively. A quantitative measure of this enhancement is useful in discussing the acceptance of such systems for diffuse (i.e., totally isotropic) radiation. Diffuse radiation equipopulates phase space. Hence it is uniformly distributed in the *L*, *M* plane. The area of the acceptance figure for an ordinary 2D CPC (corresponding to Fig. 5.3a) is $\pi\sin\theta_i$. Therefore, the fraction of diffuse radiation accepted is just $\sin\theta_i$. However, for the dielectric-filled case, the area depicted by Fig. 5.3b is found to be

$$2[n\sin\theta_i\tan^{-1}(\tan\theta_c\cos\theta_i)+\tan^{-1}(\tan\theta_i\cos\theta_c)] \tag{5.4}$$

where θ_c is the critical angle $\sin^{-1}(1/n)$. This exceeds $\pi\sin\theta_i$ by a factor

$$\frac{2}{n}[n\tan^{-1}(\tan\theta_c\cos\theta_i)+\tan^{-1}(\tan\theta_i\cos\theta_c)/\sin\theta_i] \tag{5.5}$$

This enhancement factor assumes the limiting value for small angles θ_i

$$\frac{2}{\pi}(n\theta_c+\cos\theta_c) \tag{5.6}$$

and slowly decreases to unity as θ_i increases to $\pi/2$. For example, for $n = 1.5$, a value typical for plastics, the enhancement is ≈ 1.17 for small $\theta_i(\lesssim 10°)$ and reduces to ≈ 1.13 for $\theta_i = 40°$. We shall return to this property in Chapter 10, where this extra angular acceptance is shown to be advantageous for solar energy collection.

The dielectric-filled CPC has certain practical advantages. Total internal reflection is 100% efficient, whereas it is difficult to get more than about 90% reflectivity from metallized surfaces. Also, for the same overall length the collecting angle in air is larger by the factor *n*, since Eqs. (4.1)–(4.4) for the shape of the CPC would be applied with the internal maximum angle θ_i', instead of θ_i. If the absorber can be placed into optical contact with the exit face and if it can utilize rays at all angles of incidence, the maximum theoretical concentration ratio becomes, from Eq. (2.13), $n^2/\sin^2\theta_i$, i.e., it is increased by the factor n^2 (or *n* for a trough concentrator). However, if the rays had to emerge into air at the exit aperture, it is clear that many would get turned back at this face by total internal reflection. Thus the CPC design needs to be modified for this case.

5.3 The CPC with Exit Angle Less Than $\pi/2$

There may well be instances such as that mentioned at the end of the last section where it is either impossible or inefficient to use rays emerging at up to $\pi/2$ from the normal to the exit aperture. The CPC designs can be modified very simply to achieve this. It would then be close to being what we have called an ideal concentrator but without maximum theoretical concentration ratio.

Let θ_i be the input collecting angle and θ_o the maximum output angle. Then an ideal concentrator of this kind would, from Eq. (2.12), have the concentration ratio

$$C(\theta_i, \theta_o) = (n_o \sin \theta_o)/(n_i \sin \theta_i) \tag{5.7}$$

for a 2D system or

$$C(\theta_i, \theta_o) = [(n_o \sin \theta_o)/(n_i \sin \theta_i)]^2 \tag{5.8}$$

for a 3D system. Following Rabl and Winston (1976), we may call this device a θ_i/θ_o transformer or concentrator. It is convenient to design the θ_i/θ_o concentrator by starting at the exit aperture and tracing rays back. As for the basic CPC, we start by considering the 2D case or the meridian rays for the 3D case. Let $QQ' = 2a'$ be the exit aperture in Fig. 5.4. We make all reversed rays leaving any point on QQ' at angle θ_o appear in the entrance aperture at angle θ_i to the axis. This is easily done by means of a cone section $Q'R$ making an angle $\frac{1}{2}(\theta_o - \theta_i)$ with the axis. Next we make all rays leaving Q at angles less than θ_o appear at the entry aperture at angle θ_i; this is done in the same way as for a CPC by a parabola RP' with focus at

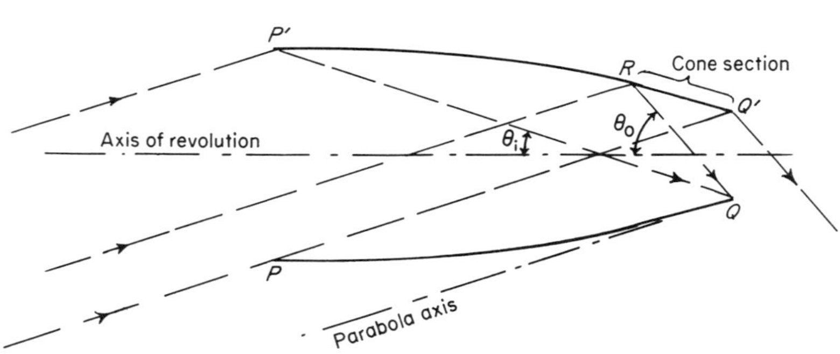

Fig. 5.4 The $\theta_i\theta_o$ concentrator; as shown, $\theta_i = 18°$ and $\theta_o = 50°$.

Q and axis at angle θ_i to the concentrator axis. The parabola finishes as usual where it meets the extreme ray from Q at θ_i, so its surface is cylindrical at the entry end.

We have ensured by construction that in the meridian section all rays entering at θ_i emerge after one reflection at less than or equal to θ_o, and it is easily seen by examining a few special cases that all rays entering at angles less than θ_i emerge at less than θ_o. This must therefore be an ideal θ_i/θ_o concentrator with

$$a/a' = \sin\theta_o/\sin\theta_i \tag{5.9}$$

We can also prove this slightly more laboriously from the geometry of the design; this is done in Appendix E.

Clearly the 2D θ_i/θ_o concentrator is an ideal concentrator since the same reasoning as in Section 4.4.3 can be applied. In 3D form there will be some losses from skew rays inside θ_i being turned back. We show in Fig. 5.5a the transmission-angle curve for a typical case. The transition is as sharp as for a full CPC, but, as can be seen by comparing Fig. 5.5b with Fig. 4.14, the pattern of rejected rays is quite different.

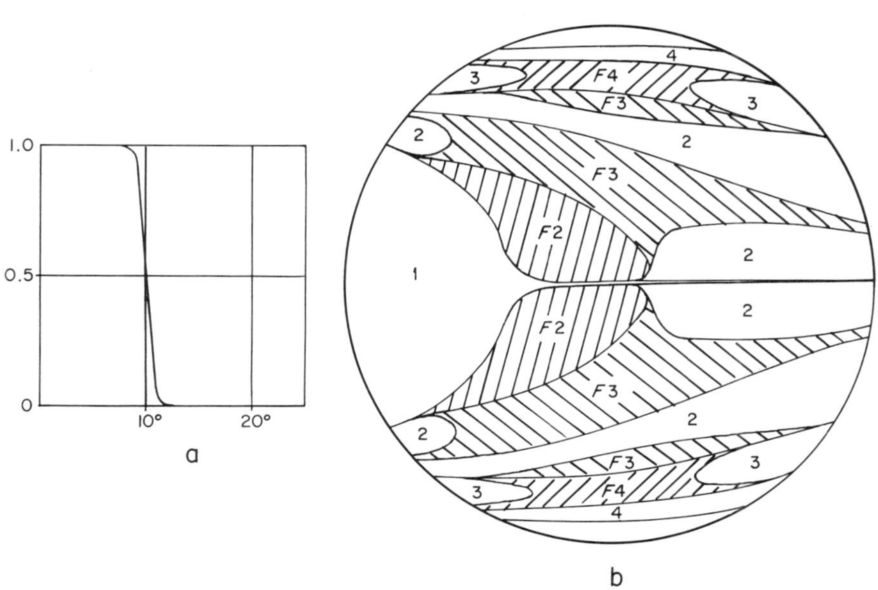

Fig. 5.5 (a) The transmission-angle curve for a 10°/60° concentrator. (b) Rays transmitted and turned back in the 10°/60° concentrator for 10° input angle (see Fig. 4.14 for full legend).

Although the 2D θ_i/θ_o concentrator transmits within θ_o all rays incident inside θ_i, it does not reject *all* rays incident outside θ_i, but transmits some of them outside θ_o. The reason for this can be seen by noting that the straight cone section in Fig. 5.4 lies outside the continuation of the parabolic section, and this parabolic section is the same as the section of an ordinary CPC with entry angle θ_i. Thus some rays outside θ_i must be transmitted and since the concentrator is ideal for rays inside θ_i, these extra rays must emerge outside θ_o. The same applies to the 3D version and the effect of these extra rays is not shown in Fig. 5.5a.

5.4 The Concentrator for a Source at a Finite Distance

So far we have assumed the source to be at infinity, as in the straightforward application to solar energy collection. There are, however, obviously cases where we should like to collect from a source at a finite distance and the edge-ray principle enables us to derive the shape very simply (Winston, 1976c).

In Fig. 5.6, let AA' be the finite source and let QQ' be the desired position of the absorber. Then if we apply the edge-ray principle, it is clear that the reflecting surface has the cross section of an ellipse, $P'Q'$, with foci at A and Q. For a 3D system the complete surface would be obtained by rotating this ellipse about the axis of symmetry.

We can show that as a 2D system this has maximum theoretical concentration by noting that all rays from AA' that enter the concentrator do emerge (by the same reasoning as for the basic CPC), and

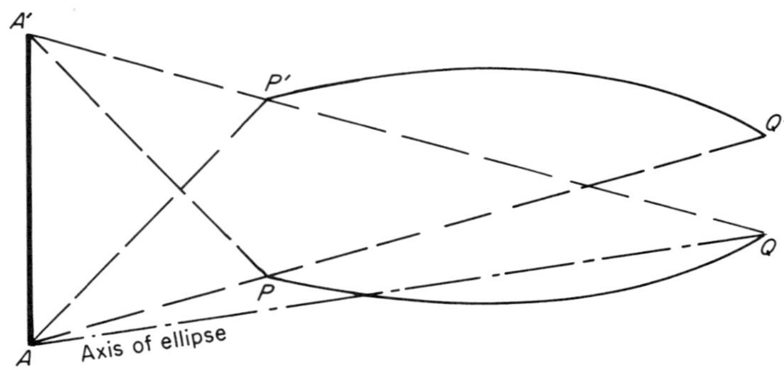

Fig. 5.6 A concentrator for the object AA' at a finite distance.

then calculating the dimensions of the system by coordinate geometry. This approach is complicated, partly because of the geometry, but also because it is not so easy to define the collecting angle for a source at a finite distance. It is better to use a more physical approach and calculate the étendue at either end of the system.

We take an object $AA' = 2\eta$ and an aperture PP' a distance z apart, as in Fig. 5.7; if the coordinate y is measured from the center of the aperture, the étendue is

$$\begin{aligned}\iint d(\sin\theta)\,dy &= \int_{-y_{\max}}^{y_{\max}} \left\{ \frac{\eta - y}{(z^2 + (\eta - y)^2)^{\frac{1}{2}}} + \frac{\eta + y}{(z^2 + (\eta + y)^2)^{\frac{1}{2}}} \right\} dy \qquad (5.10)\\ &= [-\{z^2 + (\eta - y)^2\}^{\frac{1}{2}} + \{z^2 + (\eta + y)^2\}^{\frac{1}{2}}]_{-y_{\min}}^{y_{\max}}\\ &= 2(AP' - AP)\end{aligned}$$

It is useful to rewrite Eq. (5.10) in the form

$$(A'P + AP') - (A'P' + AP) \qquad (5.11)$$

because this version, a remarkably simple result due to Hottel (1954), actually gives the correct étendue formula even when there is no particular symmetry between the relative positions of aperture and source.

Now returning to Fig. 5.6 we have, from the fundamental property of the ellipse, that the sum of the distances from the two foci to any point on the curve is a constant, i.e.,

$$AP + PQ' + Q'Q = AP' + P'Q$$

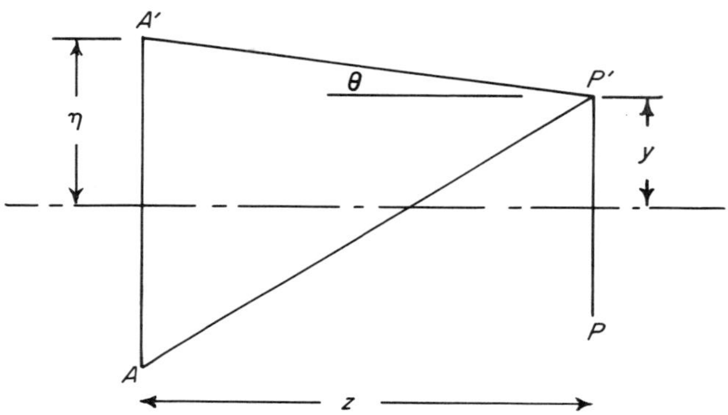

Fig. 5.7 Calculating the étendue for a source AA' at a finite distance. The collecting aperture is at PP'.

or

$$AP' - AP = Q'Q \tag{5.12}$$

so that the étendue measured at the output end is $2Q'Q$. Since $Q'Q$ is perpendicular to the axis, this must mean that rays emerge from all points of the exit aperture with their direction cosines distributed uniformly over ± 1, i.e., this system has the maximum theoretical concentration ratio.

For the 3D case with rotational symmetry (Fig. 5.6 now represents a meridional section) a straightforward calculation gives for the étendue (Winston, 1978a)

$$\pi^2(AP' - AP)^2/4 \tag{5.13}$$

while the maximum value assumed by the skew invariant is

$$h_{\max} = (AP' - AP)/2 \tag{5.14}$$

Notice that, just as for the case of an infinitely distant source ($\theta_i =$ constant), both Eqs. (5.13) and (5.14) are consistent with maximum concentration onto an exit aperture of the diameter given by Eq. (5.12). Nevertheless, this system in 3D will turn back some rays, just as for the basic CPC, and so will not be quite ideal.

5.5 The Two-Stage CPC

In Section 5.2 we described the dielectric-filled CPC using total internal reflection and we noted that the use of a refractive index greater than unity at the exit aperture permits in principle a greater concentration ratio. In order to utilize this, the absorber must be in optical contact with the dielectric, and preferably also the interface must be matched to minimize reflection losses by a suitable coating or a grading of the refractive index. This seems possible, but we then have the practical difficulty that CPCs are very long and therefore a large volume of possibly expensive dielectric is needed. This difficulty may be circumvented by a two-stage system as in Fig. 5.8. The first stage in air is a θ_i/θ_o concentrator to give whatever final output angle is needed.

This two-stage system, if designed to give maximum concentration $n/\sin\theta$, where θ is the collecting angle in air, is always longer than

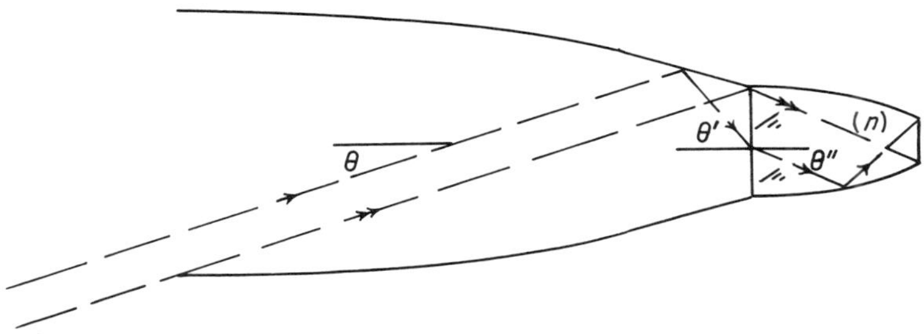

Fig. 5.8 Two-stage system ending in a dielectric of index n. In the diagram, $\theta = 10°$, $\theta' = 60°$, $\theta'' = 24.5°$, and $n = 1.85$.

a basic CPC designed for concentration $1/\sin\theta$ but with the same collecting aperture. If we make the dielectric part as short as possible, i.e., if we maximize θ' according to Eq. (5.1), then it can be shown that the overall length of the two-stage system is

$$a\left\{\cot\theta + \frac{n}{n^2-2}\cos\theta + \frac{4(n^2-1)^{3/2}}{n(n^2-2)^2}\sin\theta\right\} \quad \text{for} \quad \sqrt{2} < n \leq 2$$
$$a\{\cot\theta + \cos\theta + \tfrac{3}{2}\sqrt{3}\sin\theta\} \quad \text{for} \quad n > 2 \tag{5.15}$$

whereas the basic CPC has length

$$a(\cot\theta + \cos\theta) \tag{5.16}$$

It is easily seen that Eq. (5.15) is greater than Eq. (5.16). Figure 5.9 shows some typical values for comparison.

5.6 The CPC Designed for Skew Rays

As we saw in Chapter 4, the basic 3D CPC turns back some skew rays and this makes its concentration efficiency slightly less than ideal, as in Fig. 4.8. The basic CPC is designed by applying the edge-ray principle to meridian rays and this uses up all the degrees of freedom available, so that it is not surprising that some skew rays fail. This suggests that we should try designing a concentrator by applying the edge-ray principle to skew rays. The result might then have a slightly closer approach to maximum theoretical concentration ratio.

Figure 5.10 shows a view from the entry aperture of a concentrator. The entry aperture has diameter $2a$ and the exit aperture $2a' =$

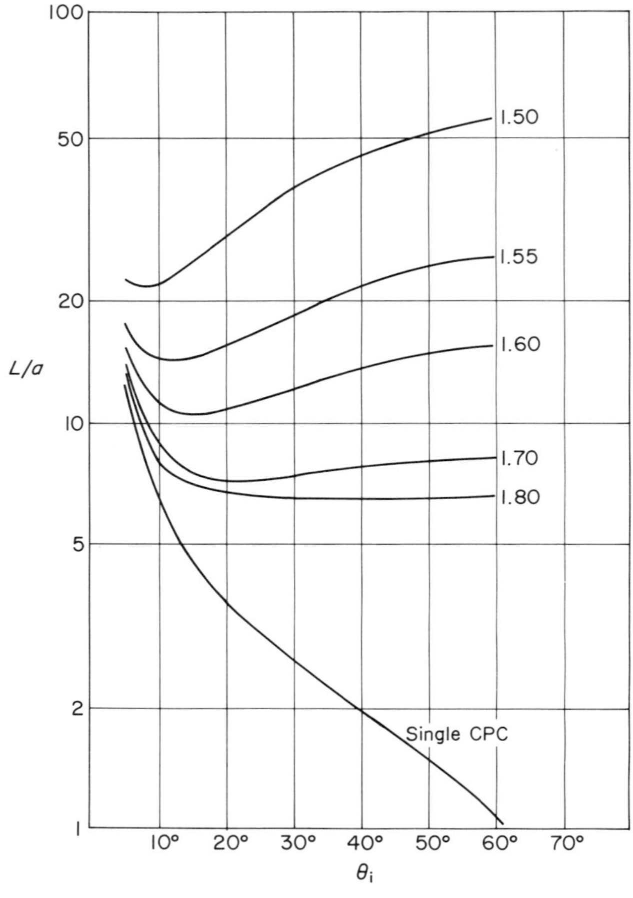

Fig. 5.9 A comparison of the overall lengths of single- and two-stage concentrators according to Eqs. (5.15) and (5.16). The ordinate is the ratio of length to radius of collecting aperture and the graphs are labeled with the refractive index of the second stage.

$2a \sin \theta_i$. Let this system be designed to fulfill the edge-ray principle for skew rays with skew invariant h (see Section 2.8 and Appendix A4 for the skew invariant). In this projected view a ray that enters at θ_i and that grazes the entry aperture at A and the exit aperture at B is reflected across to C on the exit aperture (double arrows). The

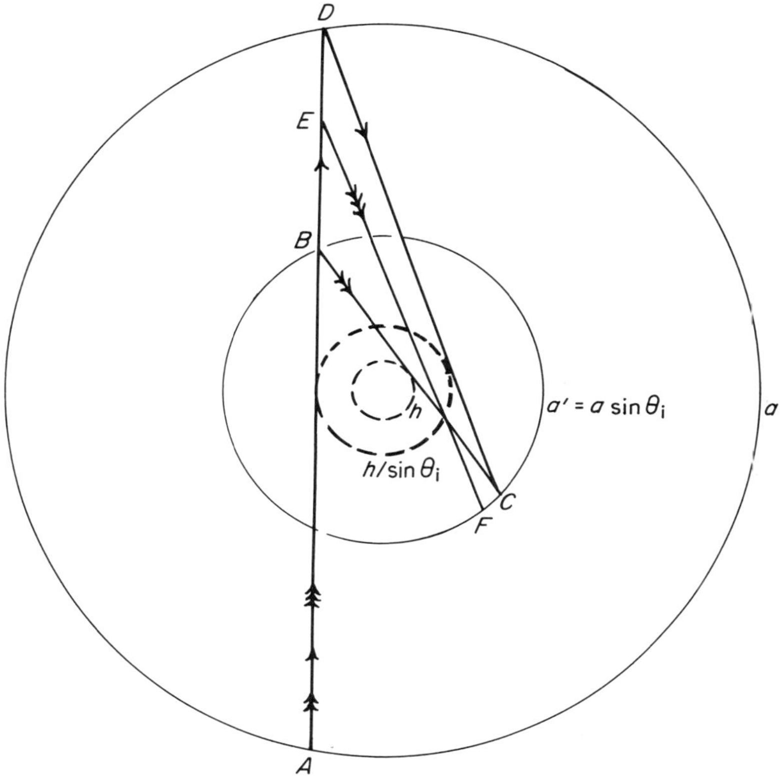

Fig. 5.10 Design of a concentrator by applying the edge-ray principle to rays with nonzero skew invariant.

segment BC is thus tangent to a circle of radius $h/\sin \theta_i$. Another ray AD, also entering at θ_i, meets the opposite entry edge at D and is reflected there. In order to satisfy the edge-ray principle, it must meet the edge of the exit aperture and therefore it is like the double-arrowed segment AB rotated around the axis so that A reaches D. Thus the segment reflected from D touches the circle of radius $h/\sin \theta_i$. We can show that this segment meets the exit aperture at the same point C as the segment BC reflected at B by calculating the angles subtended at the center by the various ray segments; the proof is given in Appendix F.

The design is completed by requiring that all rays that leave the exit aperture at its rim shall have entered at the input angle θ_i. These rays in the projection of Fig. 5.10 are typified by the three-arrowed segments, i.e., reflection at some point E along the concentrator, to emerge at F on the exit rim. It turns out that the rays do not in general emerge at the same point on the exit rim. Thus we have a situation where the edge-ray principle is less restrictive than, say, a requirement for imaging at the exit rim. This process completely determines the concentrator as a surface of revolution, but it does not seem to be possible to represent it by any analytical expression. We show in Appendix F how the differential equation of the surface, which is first-order nonlinear, is obtained. But it can only be solved numerically. P. Greenman (University of Chicago) computed the solution for several values of the input angle θ_i and skew invariant h. The results are, briefly, that the shapes are very similar to those of the basic CPC for the same θ_i, but that the overall lengths are less and the transitions in the transmission-angle curves are correspondingly more gradual. Figure 5.11 shows some of these curves.

The overall length of this concentrator is determined, as for the basic CPC, by the extreme rays, as in Fig. 5.12. This figure shows the rays ADC and ABC of Fig. 5.10, both inclined at the extreme input angle to the axis and both grazing the exit aperture after one and no reflections from the concentrator, respectively. Then in order to admit all rays at θ_i or less, the concentrator surface must finish at the point A determined by the intersection of the ray ABC with the surface. This geometry is obvious for the basic CPC, and also in that system

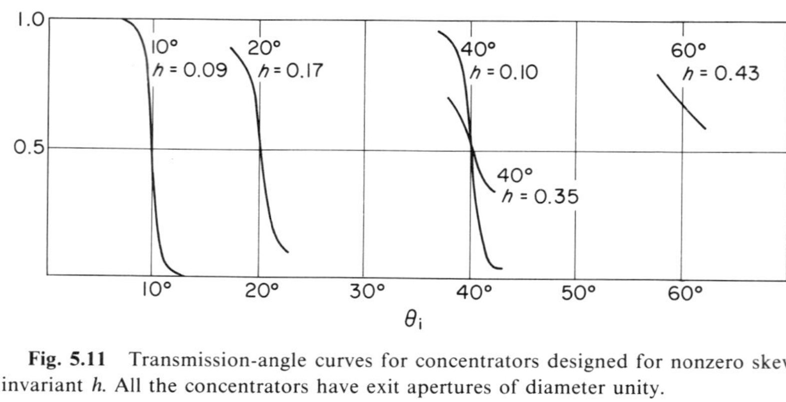

Fig. 5.11 Transmission-angle curves for concentrators designed for nonzero skew invariant h. All the concentrators have exit apertures of diameter unity.

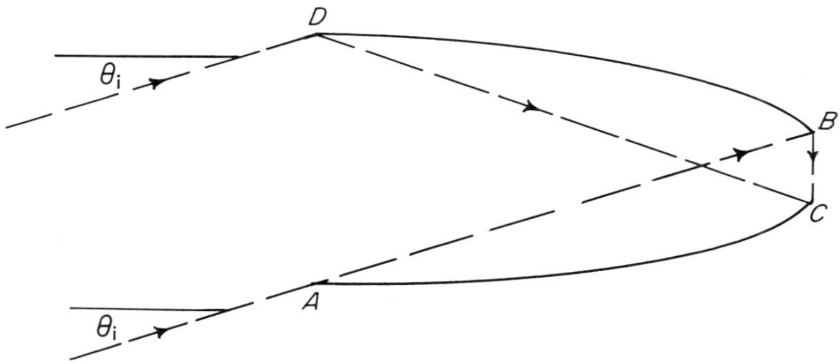

Fig. 5.12 How the length of a concentrator is determined by the extreme rays.

the ratio of input to output diameters is set as part of the design data at the desired value $1/\sin\theta_i$. In the present system the design is developed from one end and it does not follow that the ratio of input to output diameters will have any particular simple value. In fact it can be shown (Appendix F) that this ratio is again $1/\sin\theta_i$—a result that is by no means obvious. In spite of this, the concentrator with nonzero h has even less than the maximum theoretical concentration ratio than the basic CPC, as can be seen by comparing Figs. 5.11 and 4.12. This is because the shorter length permits meridian rays at angle greater than θ_i to reach the exit aperture directly. Thus, by volume conservation of phase space (Appendix A), more rays inside θ_i must be rejected.

Figure 5.13 shows a scale-drawing comparison of the basic 40° CPC with concentrators designed for nonzero h. It can be seen how meridian rays at angles greater than θ_i reach the exit aperture.

In Appendix F we show that in an ideal concentrator most transmitted rays have small values of the skew invariant h; in fact we calculate the relative frequencies of occurrence of h and show that the greatest frequency is at $h = 0$. This tends to support our finding that the solution for the concentrator design with $h = 0$ is best.

5.7 The Truncated CPC

A disadvantage of the CPC compared to systems with smaller concentration is that it is very long compared to the diameter of the

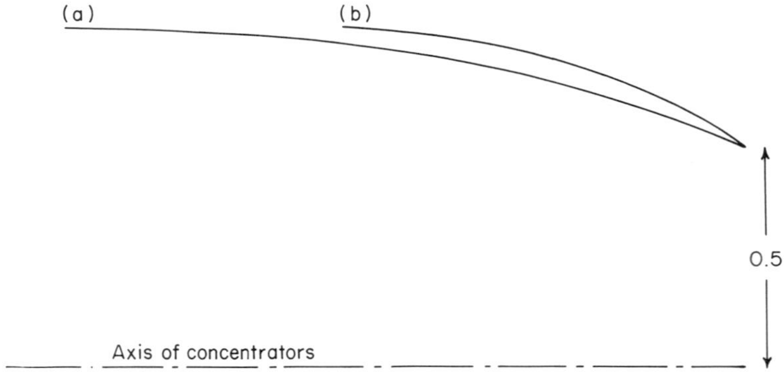

Fig. 5.13 Comparison of concentrator profiles for $\theta_i = 40°$; (a) $h = 0$; (b) $h = 0.32$. It can be seen that meridian rays from the edge of (b) at angles greater than 40° are transmitted, since rays at 40° from the edge of (a) just get through the exit aperture.

collecting aperture (or width for 2D systems). This is naturally important for economic reasons in large-scale applications such as solar energy. From Eq. (4.2) the length L is approximately equal to the diameter of the collecting aperture divided by the full collecting angle, i.e.,

$$L \sim (2a)/(2\theta_i) \tag{5.17}$$

If we truncate the CPC by removing part of the entrance aperture end, we find that a considerable reduction in length can be achieved with very little reduction in concentration, so that this may be a useful economy.

It is convenient to express the desired relationships in terms of the (r, φ) polar coordinate system as in Fig. 5.14. We denote truncated quantities by a subscript T. We are interested in the ratio of the length to the collecting apertures, and also in the ratio of the area of the reflector to that of the collecting aperture. We find

$$a_T = \frac{f \sin(\varphi_T - \theta_i)}{\sin^2 \frac{1}{2}\varphi_T} - a' \qquad f = a'(1 + \sin \theta_i) \tag{5.18}$$

$$a = \frac{a'}{\sin \theta_i}$$
$$L_T = \frac{f \cos(\varphi_T - \theta_i)}{\sin^2 \frac{1}{2}\varphi_T} \tag{5.19}$$

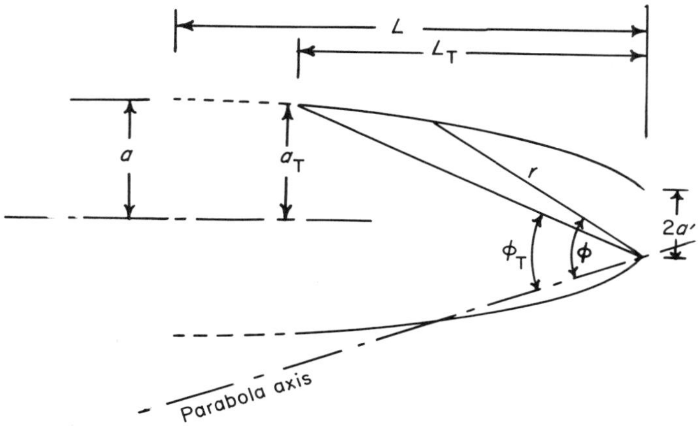

Fig. 5.14 The polar coordinates used in computing truncation effects.

$$L = f \cos \theta_i / \sin^2 \theta_i \tag{5.20}$$

so that

$$\frac{L_T}{a_T} = \frac{(1+\sin \theta_i)\cos(\varphi_T - \theta_i)}{\sin(\varphi_T - \theta_i)(1+\sin \theta_i) - \sin^2 \frac{1}{2}\varphi_T} \tag{5.21}$$

Plots of this quantity against the theoretical concentration ratio a_T/a' for 2D truncated CPCs were given by Rabl (1976c) and by Winston and Hinterberger (1975). Figure 5.15 shows some of these curves and it can be seen that initially, i.e., for points near the broken-line locus for full CPCs, the curves have a very large slope, so that the loss in concentration ratio is quite small for useful truncations.

In addition to the ratio of length to aperture diameter, we may be interested in the ratio of surface area of reflector to aperture area, since this governs the cost of material for the reflector. The general forms of the curves would be similar to those in Fig. 5.15, but with differences between 2D and 3D concentrators. The explicit formulae for reflector area divided by collector area are, for a 2D truncated CPC,

$$\frac{-f}{a_T}\left\{\frac{\cos \frac{1}{2}\varphi}{\sin^2 \frac{1}{2}\varphi} + \ln \cot \tfrac{1}{4}\varphi\right\}\Bigg|_{\varphi_T}^{\theta_i+\pi/2} \tag{5.22}$$

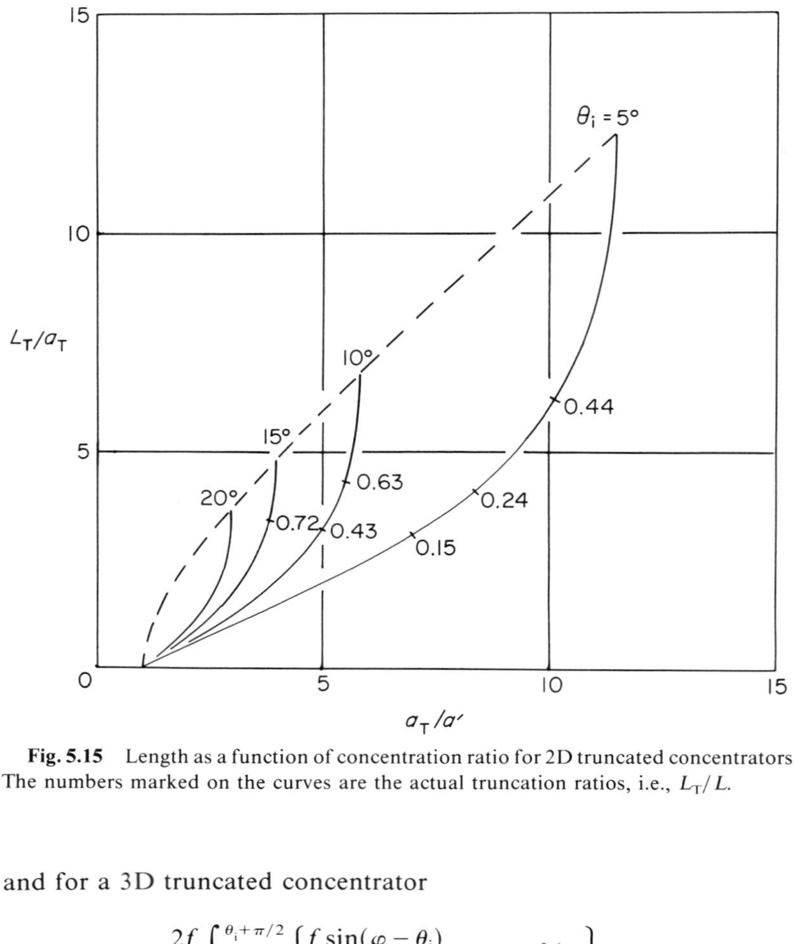

Fig. 5.15 Length as a function of concentration ratio for 2D truncated concentrators. The numbers marked on the curves are the actual truncation ratios, i.e., L_T/L.

and for a 3D truncated concentrator

$$\frac{2f}{a_T^2}\int_{\varphi_T}^{\theta_i+\pi/2}\left\{\frac{f\sin(\varphi-\theta_i)}{\sin^5\frac{1}{2}\varphi}-a'/\sin^3\tfrac{1}{2}\varphi\right\}d\varphi \tag{5.23}$$

Derivations of the above results and the explicit form of the integral in Eq. (5.23) are given in Appendix G. Some representative plots of these functions are given in Figs. 5.16 and 5.17. It should be noted that in these figures the theoretical concentration ratios are, respectively, a_T/a' and $(a_T/a')^2$. We conclude from this that the losses in performance due to moderate truncation would be acceptable in many instances on account of the economic gains.

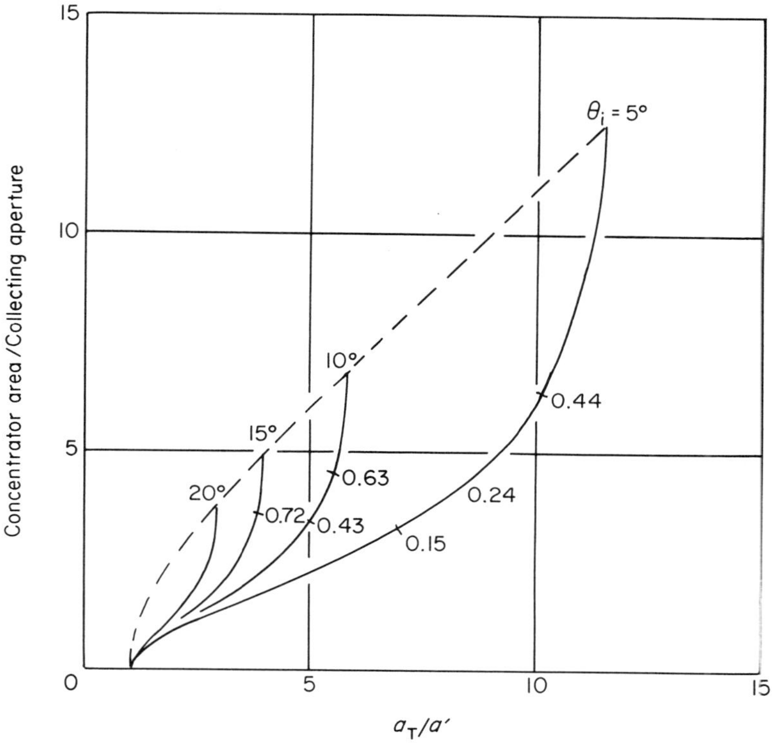

Fig. 5.16 Concentrator surface area as a function of concentration ratio for 2D truncated concentrators. The numbers marked on the curves are the actual truncation ratios, i.e., L_T/L.

5.8 The Lens–Mirror CPC

A more fundamental method for overcoming the disadvantage of excessive lengths incorporates refractive elements to converge the pencil of extreme rays. By consistent application of the edge-ray principle, we leave the optical properties of the concentrator essentially identical to the all-reflecting counterpart while substantially reducing the length in many cases. The edge-ray principle requires that the extreme incident rays at the entrance aperture also be the extreme rays at the exit aperture. In the all-reflecting construction (Fig. 4.3)

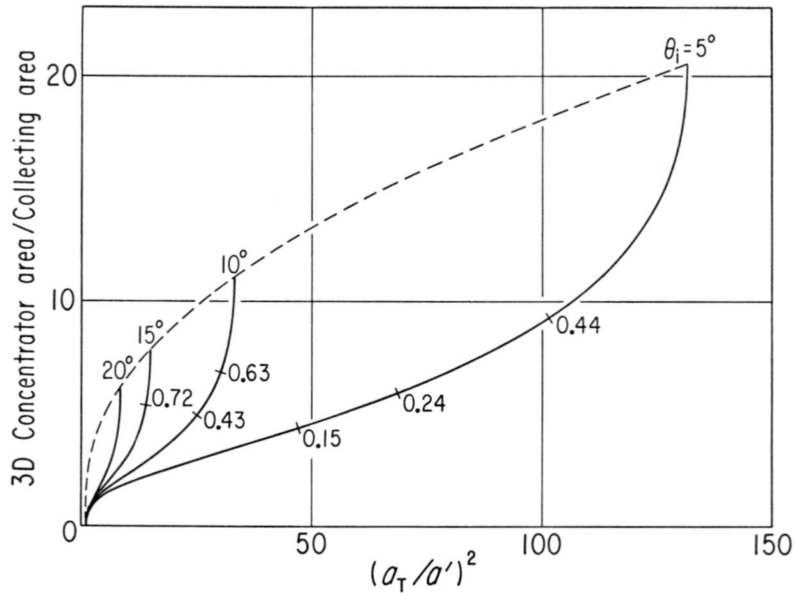

Fig. 5.17 Concentrator surface area as a function of concentration ratio for 3D truncated concentrators. The numbers marked on the curves are the actual truncation ratios, i.e., L_T/L.

this is accomplished by a parabolic mirror section that focuses the pencil of extreme rays from the wavefront W onto the point P_2 on the edge of the exit aperture. To incorporate, say, a lens at the entrance aperture the rays from W, after passage through the lens, are focused onto P_2 by an appropriately shaped mirror M (Fig. 5.18). Therefore the profile curve of M is then determined by the condition

$$\int_W^{P_2} n\, ds = \text{const.} \tag{5.24}$$

To comprehend the properties of the lens–mirror collector, it is useful to consider a hypothetical lens that focuses rays from W onto a point F. From Eq. (5.24) the appropriate profile curve for M is a hyperbola with conjugate foci at F and P_2 (Fig. 5.18).* This example illustrates the principal advantage of this configuration; the overall

* By suitable choice of parameters, the hyperbola can be a straight line.

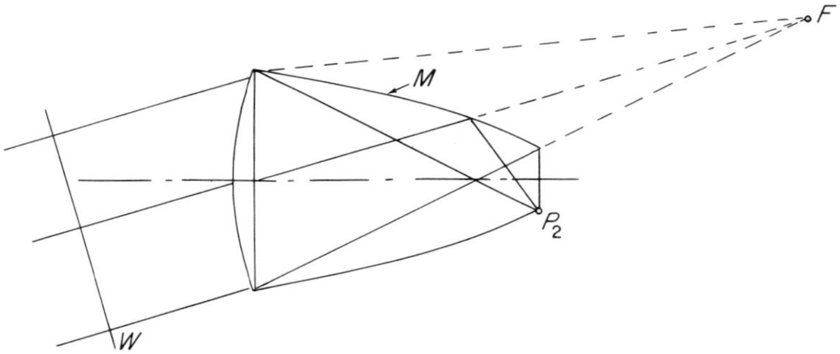

Fig. 5.18 The lens-mirror CPC.

length is greatly reduced from the all-reflecting case to $L \simeq f$, the focal length of the lens. A real lens would have comatic aberration, so that M would no longer be hyperbolic but simply a solution to Eq. (5.24). A solution will be possible so long as the aberrations are not so severe as to form a caustic between the lens and the mirror. For the example in Fig. 5.18, where the lens is plano-convex with index of refraction $n \sim 1.5$, this means we must not choose too small a value for the focal ratio of this simple lens (an f/4 choice works out nicely). Alternatively, we may say that the mirror surface corrects for lens aberrations, providing these are not too severe, to produce a sharp focus at P_2 for the extreme rays. Of course, this procedure can only be successful for monochromatic aberrations, so that it is advantageous to employ a lens material of low dispersion over the wavelength interval of interest.

We may expect the response to skew rays in a rotationally symmetric 3D system to be nonideal just as in the all-reflecting case, and, in fact, ray tracing of some sample lens-mirror configurations shows angular cut-off characteristics indistinguishable from the simple CPC counterpart. These configurations are beginning to find application to infrared collection (Keene *et al.*, 1977) and in the 2D version for solar energy concentration (Collares-Pereira *et al.*, 1977). We note that certain configurations of the lens-mirror type were proposed by Ploke (1967).

CHAPTER 6

Developments of the Compound Parabolic Concentrator for Nonplane Absorbers

6.1 2D Collection in General

For moderate concentration ratios for solar energy collection there is considerable interest in systems that do not need diurnal guiding, for obvious reasons of economy and simplicity (see, e.g., Winston, 1974; Winston and Hinterberger, 1975). These naturally would have troughlike or 2D shapes and would be set pointing south* at a suitable elevation so as to collect flux efficiently over a good proportion of the daylight hours. So far our discussion has suggested that these might take the form of 2D CPCs, truncated CPCs, or compound systems with a dielectric-filled CPC as the second stage. In discussing all these it was tacitly assumed that the absorber would present a plane surface to the concentrator at the exit aperture and this, of course, made the geometry particularly simple. In fact, when applications are considered in detail, it becomes apparent that other shapes of absorber would be useful. In particular, it is obvious that cylindrical absorbers, i.e., tubes for heating fluids, suggest themselves. In this chapter we discuss the developments in design necessary to take account of such requirements.

* In the northern hemisphere.

6.2 Extension of the Edge-Ray Principle

In Chapter 3 we proposed the edge-ray principle as a way of initiating the design of concentrators with concentration ratio approaching the maximum theoretical value. We found that for the 2D CPC this maximum theoretical value was actually attained by direct application of the principle.

We now propose a way of generalizing the principle to nonplane absorbers in 2D concentrators. Let the concentrator be as in Fig. 6.1, which shows a generalized tubular absorber. We assume the section of the absorber is convex everywhere, and we also assume it is symmetric about the horizontal axis indicated. Then we assert that the required generalization of the edge-ray principle is that rays entering at the maximum angle θ_i shall be tangent to the absorber surface after one reflection, as indicated.

The generalization can easily be seen to reduce to the edge-ray principle for a plane absorber. In order to calculate the concentration, we need to have a rule for constructing the concentrator surface beyond the point P_0' at which the extreme reflected ray meets the surface. Here we choose to continue the reflector as an involute of the absorber surface, as indicated by the broken line. A reason for this choice will

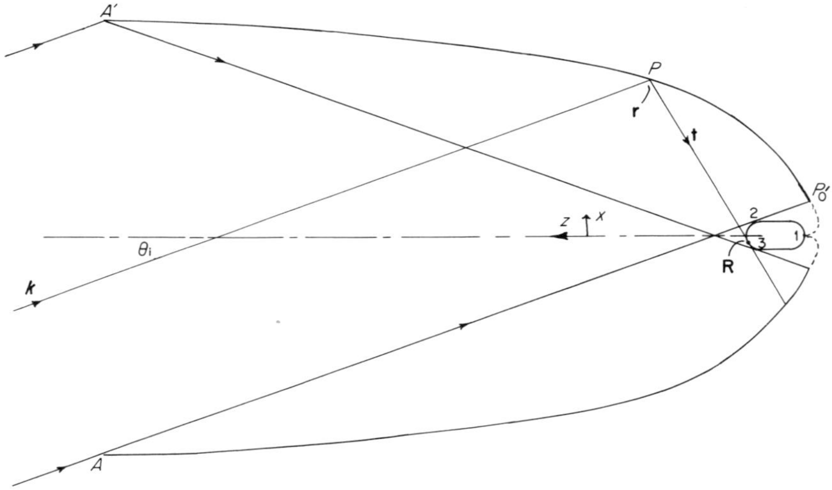

Fig. 6.1 Generalizing the edge-ray principle for a nonplane absorber.

be suggested below. We shall be able to show that this design for a 2D concentrator achieves the maximum possible concentration ratio, defined in this case as the entry aperture area divided by the area of the curved absorber surface.

Following Winston and Hinterberger (1975), we let $\mathbf{r}$ be the position vector of a current point P on the concentrator surface, and take $\mathbf{R}$ as the position vector of the point of contact of the ray with the absorber. Then we have

$$\mathbf{r} = \mathbf{R} - l\mathbf{t} \tag{6.1}$$

where $\mathbf{t}$ is the unit tangent to the absorber, i.e., we have

$$\mathbf{t} = d\mathbf{R}/dS \tag{6.2}$$

where S is the arc length round the absorber. Let $\mathbf{k}$ be a unit vector along the direction of the extreme rays, so that in the coordinate system shown $\mathbf{k} = (\sin\theta_i, 0, -\cos\theta_i)$. Our condition that the tangent to the absorber be reflected into $\mathbf{k}$ takes the form, equating sines of the angles of incidence and reflection,

$$\mathbf{t} \cdot d\mathbf{r} = \mathbf{k} \cdot d\mathbf{r}$$

or

$$\mathbf{t} \cdot \frac{d\mathbf{r}}{dS} = \mathbf{k} \cdot \frac{d\mathbf{r}}{dS} \tag{6.3}$$

Now by differentiating Eq. (6.1) we obtain

$$\frac{d\mathbf{r}}{dS} = \frac{d\mathbf{R}}{dS} - \frac{dl}{dS}\mathbf{t} - \frac{l\,d\mathbf{t}}{dS}$$

and on scalar multiplication by $\mathbf{t}$ this gives

$$\mathbf{t} \cdot \frac{d\mathbf{r}}{dS} = 1 - \frac{dl}{dS} \tag{6.4}$$

On substituting in Eq. (6.3) and integrating we obtain

$$(S - l)\Big|_2^3 = (\mathbf{r}_3 - \mathbf{r}_2) \cdot \mathbf{k} \tag{6.5}$$

In this equation the points 2 and 3 would be those corresponding to the extreme reflected rays, as in the diagram.

Between points 1 and 2 we have postulated that the concentrator profile shall be an involute of the absorber, and the condition for this is

$$\mathbf{t}\cdot\frac{d\mathbf{r}}{dS}=0 \tag{6.6}$$

Thus for this section of the curve we have from Eq. (6.4)

$$S_2-S_1=l_2-l_1$$

or, since our involute is chosen to be the one that starts at point 1,

$$S_2=l_2 \tag{6.7}$$

Thus Eq. (6.5) gives

$$S_3-l_3-(\mathbf{r}_3-\mathbf{r}_2)\cdot\mathbf{k} \tag{6.8}$$

From the figure it can be seen that $(\mathbf{r}_3-\mathbf{r}_2)\cdot\mathbf{k}$ is equal to the projection of $P_0'A'$ onto AP_0'. Thus

$$(\mathbf{r}_3-\mathbf{r}_2)\cdot\mathbf{k}=-(l_3+l_2)+2a\sin\theta_\mathrm{i}$$

and on substituting into Eq. (6.8) we find

$$S_3+l_2=2a\sin\theta_\mathrm{i} \tag{6.9}$$

Recalling that the second section of the concentrator is an involute, we see that $l_2=S_2$ and thus

$$S=S_3+S_2=2a\sin\theta_\mathrm{i} \tag{6.10}$$

We have proved that the concentrator profile generated in this way has the theoretical ratio of input area to absorber area, i.e., it has the maximum theoretical concentration ratio if no rays are turned back.

If the property of the involute that its normal is tangent to the parent curve is remembered, it is easy to see that a concentrator designed in this way sends all rays inside the angle θ_i to the absorber, including those outside the plane of the diagram if it is a 2D system. Thus from arguments based on étendue and on phase space conservation (see Section 2.7 and Appendix A), the system is optimal.

6.3 Some Examples

It is easy to apply our generalization to plane absorbers. Figure 6.2 shows an edge-on fin absorber QQ' with extreme rays AQP_0' and

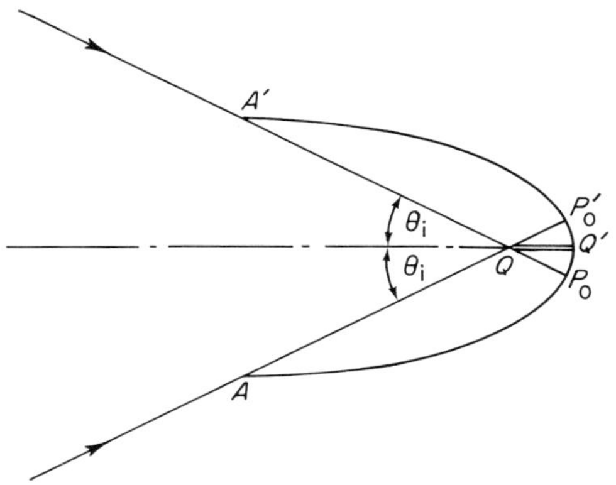

Fig. 6.2 The optimum concentrator design for an edge-on fin.

$A'QP_0$. Clearly, the section of the concentrator between P_0 and P_0' is an arc of a circle centered on Q and the section $A'P_0'$ is a parabola with focus at Q and axis AQP_0'.

The two-sided flat-plate collector normal to the axis, as in Fig. 6.3, is a slightly more complicated case. Following our rules there are three sections to the profile. OP' receives no direct illumination and is thus an involute of the segment OQ', i.e., an arc of a circle centered on Q'. $P'R'$ must focus extreme rays on Q' and it is therefore part of a parabola with focus at Q and axis $AQ'P'$. $R'A'$ must focus extreme rays on Q and is therefore a parabola with focus Q and axis parallel to $AQ'P'$.

In all cases it can easily be seen from the general mode of construction described in Section 6.2 that the segments of different curves have the same slope where they join; for example, in Fig. 6.3 the normal at P' is a ray for the circular segment OP' and the parabolic segment $P'R'$, and at R' the incident ray at angle θ_i is required to be reflected to Q' by the segment $P'R'$ and to Q by the segment $R'A'$.

Figure 6.4 shows to scale the profile for a circular section absorber. Here the actual profile does not have a simple parabolic or circular shape, but we shall give the solution in Section 6.4. It is noteworthy, however, that in Section 6.2 we deduced the property of having maximum theoretical concentration ratio without explicit reference to the profile, just as we were able to do for the basic CPC (see Chapter 4). The case of the circular section absorber is important for solar

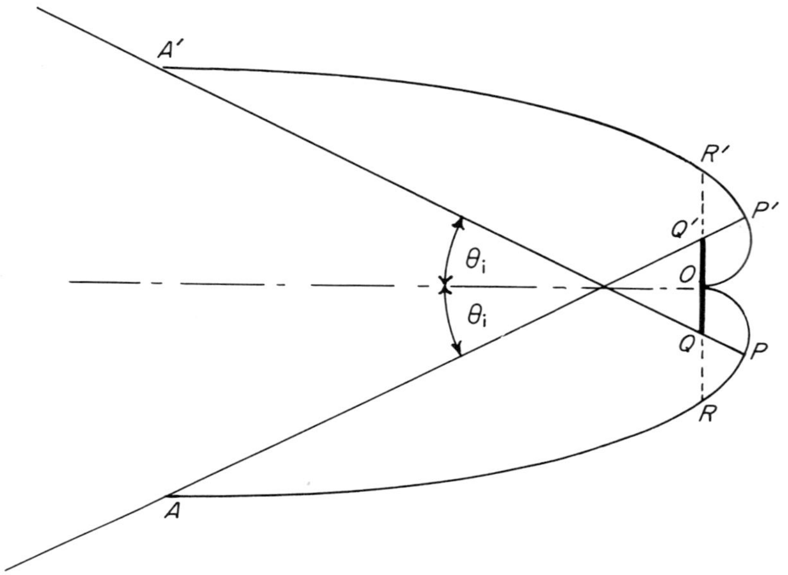

Fig. 6.3 The optimum concentrator for a transverse fin.

energy applications (Chapter 10) and has been actively pursued by a number of investigators. For example, Ortobasi (1974) independently developed the ideal mirror profile in an innovative collector program at Corning Glass Company..

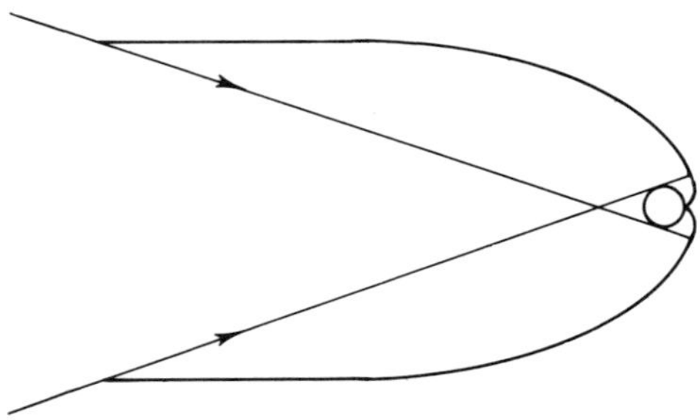

Fig. 6.4 The optimum concentrator for a cylindrical absorber.

6.4 The Differential Equation for the Concentrator Profile

It is straightforward but laborious to set up a differential equation for the concentrator profile. The equation, given with its solution in Appendix H, is used for the region of the profile that sends extreme rays tangent to the absorber after one reflection, i.e., the region between points 2 and 3 in Fig. 6.1. The remaining region is an involute arranged to join the profile smoothly, and the equation for this is also given in the appendix. However, it is worth noting that for many practical applications the involute curve can be drawn accurately enough to scale by the draftsman's method of unwinding a taut thread from the absorber profile.

6.5 Mechanical Construction for 2D Concentrator Profiles

In the discussion of Fig. 6.1 in Section 6.2 it appeared that part of the concentrator surface was generated as an involute of the absorber section. It is possible to combine this result with the fact that optical path lengths from a wave front to a focus are constant to obtain a simple geometrical construction for the concentrator profile. Figure 6.5 shows a system similar to that shown in Fig. 6.1, but we have

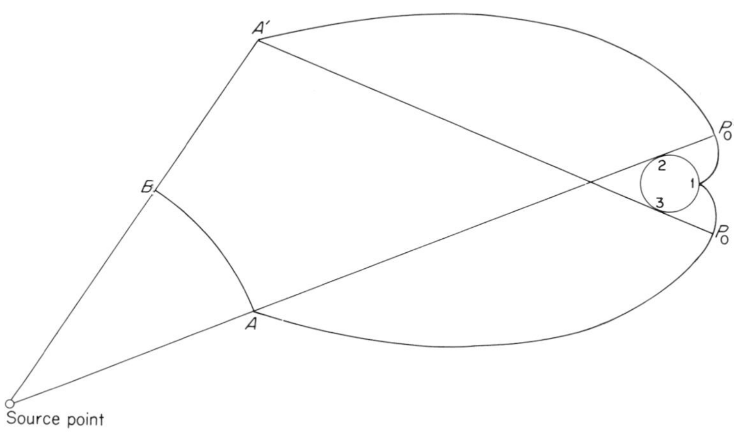

Fig. 6.5 A concentrator for a source at a finite distance and a nonplane absorber.

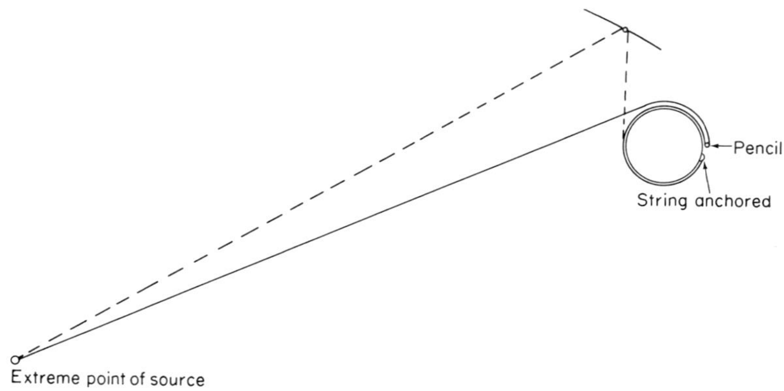

Fig. 6.6 The string construction for the concentrator profile.

drawn in a wave front AB of the incoming extreme pencil and we assume the source is at a large but finite distance.

The construction is then as follows. We tie a string between the source and point 1 at the rear of the absorber and we pull the string taut with a pencil, as in the so-called gardener's method of drawing an ellipse. The length of the string must be such that it will be just taut when it is pulled right around the absorber to reach point 1 from the other side, as in Fig. 6.6. It is then unwound, keeping the string taut, and the pencil describes the correct profile. To check this, we simply have to show that the line drawn is at the correct angle to produce reflection. In Fig. 6.7 the string is tangent to the absorber at A, the source point is at C, and B is a typical position of the pencil. If the pencil is moved to B' where BCB' is a small angle ε, then we have

$$CBA = CB'A' - AA' + 0(\varepsilon^2) \tag{6.11}$$

so that

$$CBA = CB'A + 0(\varepsilon^2) \tag{6.12}$$

Thus by Fermat's principle BB' must be a portion of a reflecting surface that reflects CB into BA.

Figure 6.8 shows this method generalized further to a convex source and a convex absorber. The string is anchored at two suitably chosen points A and B and stretched with the pencil P. The length of the string is chosen so that it just reaches to the point Q when wound

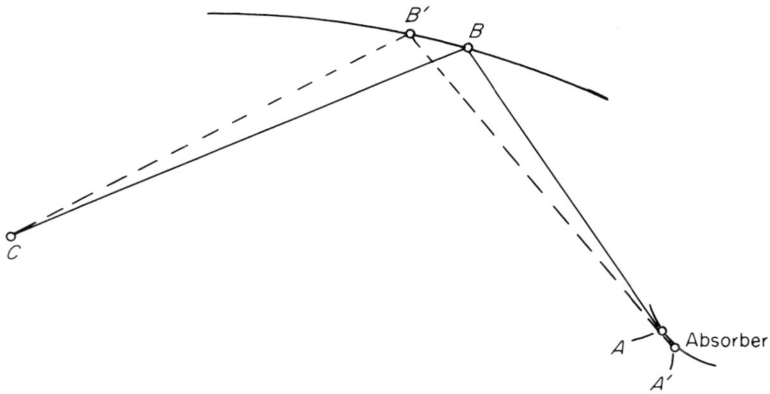

Fig. 6.7 Proof that the string construction gives a concentrator surface that agrees with Fermat's principle.

round the absorber. It is easily seen that this generates a 2D ideal concentrator. Here we are generating a concentrator that collects all the flux from the source and sends it all to the absorber, and for this to be physically possible we must make the perimeters of the source and absorber equal. If this were not so, the construction would not

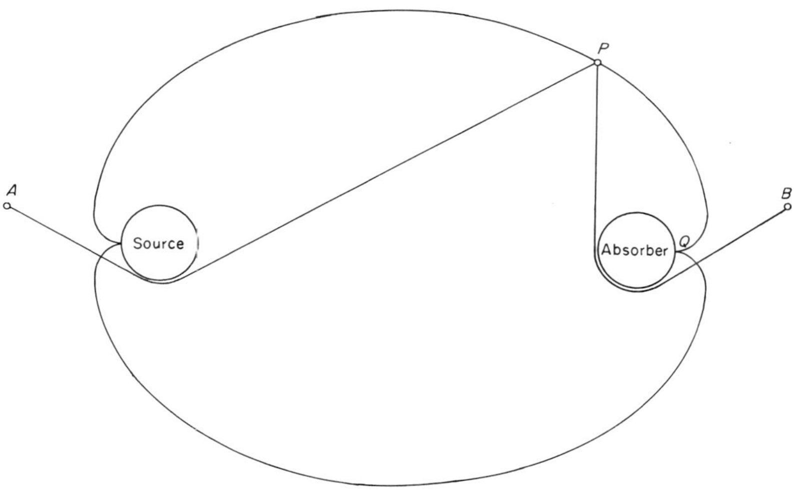

Fig. 6.8 The case of a convex source and a convex absorber treated by the string construction.

work, in the sense that the string would not be of the right length to just close the curve, and this would mean that we were trying to infringe the rules dictated by conservation of étendue.

Returning to Fig. 6.8, we note that a solution is possible for any distance between source and absorber. We note also that the solution appears not to be unique, in the sense that we could break the reflector at its widest part and insert a straight parallel-sided section of any length, since such a section clearly transforms an étendue of width $2a$ and angle π. However, from obvious practical considerations, it is desirable to minimize the number of reflections of a given ray and this is clearly done by not inserting such a straight section.

We could, of course, vary these solutions still further by adding in CPCs and θ_o/θ_i systems (Section 5.3) at the above break points instead of or in addition to the cylindrical sections. Again, this adds to the number of reflections, and it seems that there is always a unique "most economical" solution for which the extreme rays have at most one reflection.

We could generalize the string method to media with nonuniform refractive index by postulating that the string is elastic in such a way that it always assumes the optical path length, $\int n\, ds$, of the medium through which it passes. There is, of course, no way in which such a string could be realized, but the concept shows how ray trajectories, i.e., geodesics, between the extremes of the apertures always define the correct mirror surface.

In another example of parallel development, the "string method" for generating ideal mirror profiles was discovered independently by Bassett and Derrick (1978) of the University of Sydney.

6.6 The Most General Design Method for a 2D Concentrator

In this section we formulate a very general treatment of the 2D concentrator, from which most other results in this chapter could be derived. We shall describe a procedure in which an input surface and an output surface of given shapes are postulated enclosing and surrounded by regions of given refractive index distributions. On each of these surfaces a distribution of extreme rays is given. Then the procedure enables us to design a 2D concentrator that will ensure that

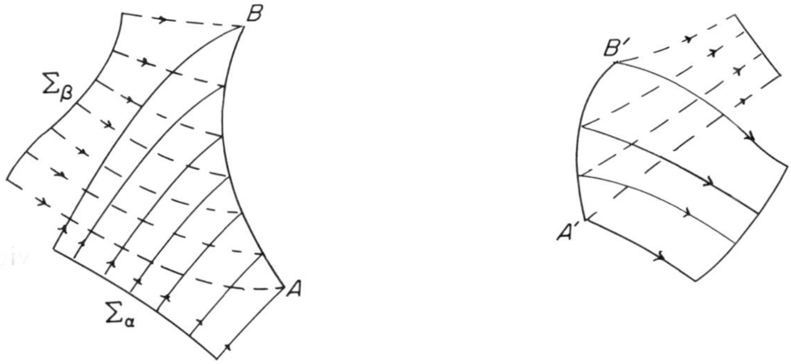

Fig. 6.9 Beams of equal étendue to fit a concentrator.

all rays between the extreme incoming rays and none outside are transmitted, so that the concentrator is optimal.

Suppose we have, as in Fig. 6.9, two surfaces AB and $A'B'$ and let AB be illuminated in such a way that the extreme angle rays at each point form pencils belonging, respectively, to wave fronts Σ_α and Σ_β. Similarly, rays at intermediate angles belong to other wave fronts, so that the whole ensemble of rays comes ultimately from a line of point sources and is transformed by a possibly inhomogeneous medium in such a way that the rays just fill the aperture AB. These rays then have a certain étendue H and we shall see below how to calculate it. Similarly, we draw rays and wave fronts emerging from $A'B'$ as indicated and we postulate that these shall have the same étendue H.

Now we ask How can we design a concentrator system between the surfaces AB and $A'B'$, possibly containing an inhomogeneous medium, that shall transform the incoming beam into the emergent beam without loss of étendue?

To solve this problem, we postulate a new principle (we shall see that our edge-ray principle of Chapter 4 can be regarded as derived from it): The optical system between AB and $A'B'$ must be such as to exactly image the pencil from the wave front Σ_α into one of the emergent wave fronts and Σ_β into the other. By "exactly" we mean that all rays from Σ_α as delimited by the aperture AB must just fill $A'B'$ so that none is lost and there is no unused space, and similarly for the rays from Σ_β.

At this point it may be objected that the above seems to have little connection with our original edge-ray principle. But consider a system

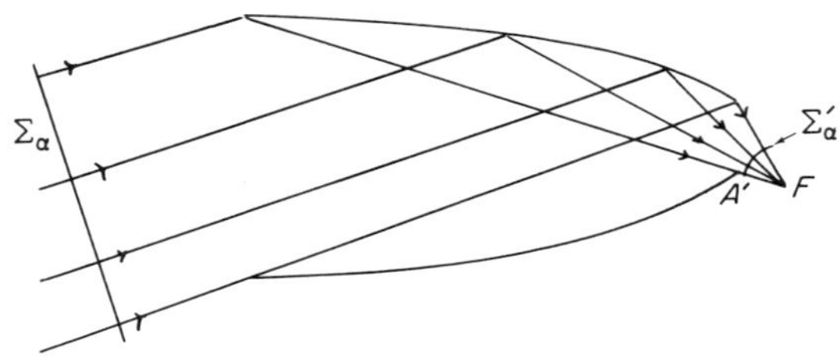

Fig. 6.10 The edge-ray principle as a limiting case of matching wave fronts.

such as in Fig. 6.10, which shows rays from one extreme wave front Σ_α in a CPC-like concentrator, and also a wave front Σ'_α. Clearly, the above principle is satisfied for these two. Now suppose the system is modified gradually so that the focus F of the emergent wave front Σ'_α gradually moves into coincidence with the edge A' of the exit aperture. We then recover the CPC geometry and the original edge-ray principle. Thus our new principle could be stated in the form that extreme points of the source must be imaged through the system by rays that just fill the exit and entry apertures.

To see how this principle leads to a solution of the general problem stated at the beginning of this section, we must first show how to calculate the étendue of an arbitrary beam of rays at a curved aperture, as in Fig. 6.9. We use the Hilbert integral, a concept from the calculus of variations. In the optics context (Luneburg, 1964), the Hilbert integral for a path from P_1 to P_2 across a pencil of rays that originated in a single point is

$$I(1,2) = \int_{P_1}^{P_2} n\mathbf{k} \cdot d\mathbf{s} \tag{6.13}$$

where n is the local refractive index, $\mathbf{k}$ is a unit vector along the ray direction at the current point, and $d\mathbf{s}$ is an element along the path P_1P_2. Thus $I(1,2)$ is simply the optical path length along any ray between the wave fronts that pass through P_1 and P_2, so that it is independent of the form of the path of integration. We can now use this to find the étendue of the beams in Fig. 6.11. The Hilbert integral from A to B for the α pencil is seen from Eq. (6.14) to be

$$I_\alpha(AB) = \int_A^B n \sin\varphi \, ds \tag{6.14}$$

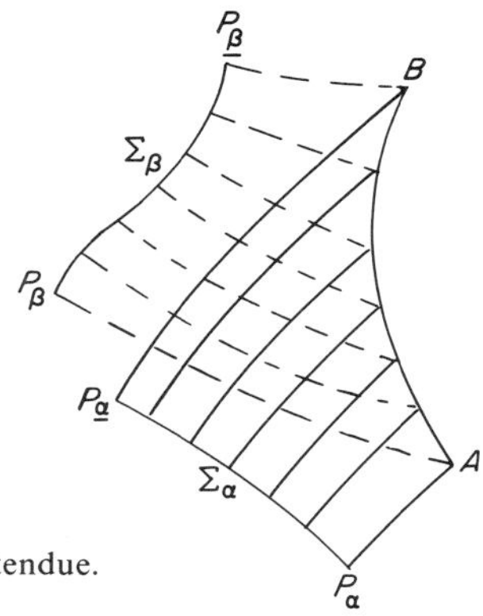

Fig. 6.11 Calculating the étendue.

where φ is the angle of incidence of a ray on the line element ds. Thus

$$I_\alpha(AB) = \langle n \sin \varphi \rangle L_{AB} \tag{6.15}$$

where $\langle\ \rangle$ denotes the average and L_{AB} is the length of the curve from A to B.

It follows that provided the aperture AB is filled with rays from the line of point sources mentioned above, the étendue is simply

$$I_\alpha(AB) - I_\beta(AB)$$

But

$$I_\alpha(AB) = [P_{\underline{\alpha}} B] - [P_\alpha A] \tag{6.16}$$

where the square brackets denote optical path lengths along the rays, so that we obtain for the étendue

$$H = [P_{\underline{\alpha}} B] + [P_\beta A] - [P_\alpha A] - [P_{\underline{\beta}} B] \tag{6.17}$$

This result can be seen to be a simple generalization of the result of Eq. (5.11).

Now let there be some kind of system constructed that achieves the desired transformation of incident extreme pencils with emergent extreme pencils, as in Fig. 6.12. The system takes Σ_α into Σ'_α and Σ_β into Σ'_β and we wish it to do so without loss of étendue. We write down the optical path length from$P_{\underline{\alpha}}$ to $P'_{\underline{\alpha}}$ and equate it to that from P_α to P'_α. Similarly, for the other pencil,

$$\begin{aligned} [P_{\underline{\alpha}} B] + [BA']_\alpha + [A'P'_{\underline{\alpha}}] &= [P_\alpha A] + [AB']_\alpha + [B'P'_\alpha] \\ [P_\beta A] + [AB']_\beta + [B'P'_\beta] &= [P_{\underline{\beta}} B] + [BA']_\beta + [A'P'_{\underline{\beta}}] \end{aligned} \tag{6.18}$$

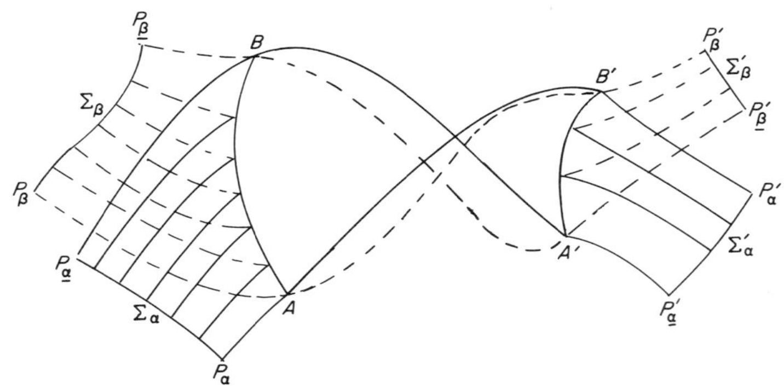

Fig. 6.12 Rays inside the concentrator.

where $[BA']_\beta$ denotes the optical path from B to A' along the β ray and similarly for the other symbols. From these we find

$$\begin{aligned}&\{[P_{\underline{\alpha}} B]+[P_\beta A]-[P_\alpha A]-[P_{\underline{\beta}} B]\}\\&\quad-\{[A'P'_{\underline{\beta}}]+[B'P'_\alpha]-[A'P'_{\underline{\alpha}}]-[B'P'_\beta]\}\\&\qquad=[AB']_\alpha-[AB']_\beta+[BA']_\beta-[BA']_\alpha\end{aligned} \tag{6.19}$$

The left-hand side of this equation can be seen by comparison with Eq. (6.17) to be the difference between the étendues at the entry and exit apertures. Since we require this difference to vanish, we have to make the right-hand side of Eq. (6.19) vanish. A simple way to do this would be to ensure that the optical system of the concentrator is such that the α and β ray paths from A to B' coincide, and similarly those from B to A'. We can do this by starting segments of mirror surfaces at A and A' in such directions as to bisect the angles between the incoming α and β rays. We then continue the mirror surfaces in such a way as to make all β rays join up with the corresponding emerging β' rays, i.e., we image the β pencil exactly into the β' pencil, and similarly for the other mirror surface connecting B and B'. We have thus completed the construction and used up all degrees of freedom in doing so.

It appears from this theorem that it is always necessary to incorporate mirrors in the design of ideal concentrators. Something like this emerges in considering simpler symmetric concentrators based on the edge-ray principle. The concentrator has to image the extreme ray pencils sharply at the rim of the exit aperture, and since any lens extends equally on either side of the axis, it must operate on both extreme ray pencils. This requires that a given region of the lens away from the axis shall produce stigmatic imagery of two pencils at different angles, which seems to be impossible for a finite number of lens elements except for the rather special cases mentioned in Appendix B.

6.7 A Constructive Design Principle for Optimal Concentrators

We conclude this chapter with a discussion of a design prescription for optimal concentrators that historically has been a fertile source of new and useful solutions. In contrast to the edge-ray principle, which is allied to such abstract notions as étendue and the Hilbert integral theorem, this practical procedure directly instructs us how to draw the profile curve of the mirror. The statement is simply that we maximize the slope of the mirror profile curve consistent with reflecting the extreme entrance rays onto the absorber and subject to various subsidiary conditions which we may wish to impose. This generic principle for designing optimal concentrators is inherent in the earliest references (Hinterberger and Winston, 1966a).

To see how this rule operates, we notice that the CPC designs for variously shaped absorbers are obtained when we don't impose any subsidiary conditions. On the other hand, the θ_i/θ_o design results when we impose the condition that the maximum exit angle not exceed θ_o. This approach is particularly useful in situations in which the subsidiary conditions preclude an ideal system, hence strict application of the edge-ray principle. For example, the various totally internally reflecting designs are obtained by specifying total internal reflection at the external wall as a condition. As discussed in Chapter 5, when the input ray distribution and index of refraction fall outside a specified range, ideal solutions are not possible. However, efficient designs are still available.

Consider the example in Fig. 6.13, where the front entrance face is curved and the index of refraction is not sufficiently high to totally reflect the extreme rays along the portion P_2, P_3 were we to follow the edge-ray principle. The maximum slope rule would suggest we maintain a constant angle (the critical angle) between the extreme ray

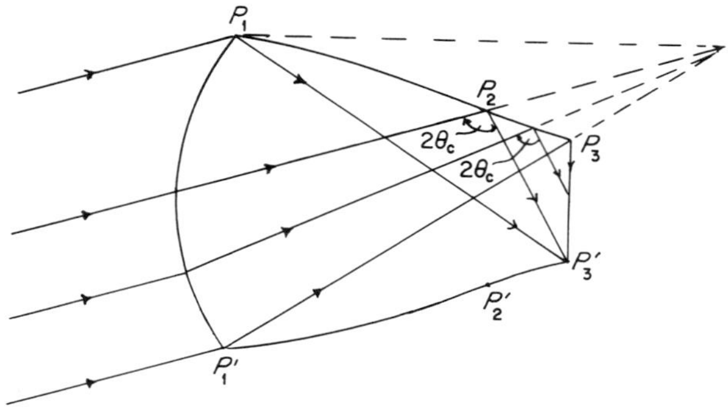

Fig. 6.13 The solid concentrator that just satisfies the critical angle condition. The profile is hyperbolic from P_1 to P_2 (assuming the entry surface is aberration free). Then if the critical angle θ_c for total internal reflection is reached at P_2, it is an equiangular spiral.

and normal to the wall along this portion of the profile curve. In the approximation that the curved face images the extreme pencil onto a point, the portion between P_1, P_2 is a hyperbola while the portion between P_2, P_3 is an arc of an equiangular spiral.

CHAPTER 7

Flow Line Approach to Nonimaging Concentration

7.1 The Concept of the Flow Line

We have seen that the edge-ray approach to nonimaging concentration has considerable flexibility in handling 2D problems. Its application in 3D is more limited and, while useful designs are obtainable in 3D, these fall short of ideal concentration. This motivates the search for alternatives to edge-ray designs, especially in 3D applications (Winston and Welford, 1979a).

In the geometrical optics approximation, we describe the propagation of light rays as trajectories in a six-dimensional phase space (p, q), where the components of q are the generalized coordinates (q_1, q_2, q_3) and the components of p are the generalized momenta. These momenta give the optical path length when the light ray traverses length ds through a medium with an index of refraction n by the relation $n\,ds = p_1 dq_1 + p_2 dq_2 + p_3 dq_3$. For example, in cartesian coordinates (x, y, z) the components of p are just the optical-direction cosines.

It is useful to picture the propagation of the totality of light rays as if it were a fluid flow in this six-dimensional space. This flow is subject to conservation theorems which are essentially geometric and follow from the law of propagation of light rays. These are loosely referred to as Liouville's theorem but are more properly called the integral invariants of Poincaré. For our purpose, the most useful invariant is constructed out of a vector **J**, whose components are

$$\mathbf{J}_1 = \int dp_2\, dp_3, \qquad \mathbf{J}_2 = \int dp_1\, dp_3, \qquad \mathbf{J}_1 = \int dp_1\, dp_2 \tag{7.1}$$

Postulating as always a medium with no attenuation or radiation sources, the surface integral of $\mathbf{J} = \int \mathbf{J} \cdot dA$ is invariant so that **J** has zero divergence. We call **J** the geometrical vector flux and the direction of **J** the flow line. Quantities like **J** occur in radiative transfer where $\mathbf{J} \cdot d\mathbf{A}$ is proportional to the net flux through the surface element $d\mathbf{A}$ and in photometry where $\mathbf{J}_z$ is proportional to the illumination on the x, y plane. This is most readily seen by rewriting

$$\mathbf{J}_z = \int n^2\, d\Omega \cos\theta z \tag{7.2}$$

where θ_z is the inclination of the elementary beam to the z axis.

7.2 Lines of Flow from Lambertian Radiators: 2D Examples

All Lambertian radiators which subtend the same solid angle at a point in the strict sense that they are bounded by the same cone give the same vector flux **J** at that point. This can be made clear by some simple examples (see Winston and Welford, 1979b).

Consider a strip AA' in the x, y plane, of width 2 in the y direction, as in Fig. 7.1. A simple calculation shows that **J** points along the bisector of the angle APA' and its magnitude is $|\mathbf{J}| = 2 \sin\theta$ where the angle $APA' = 2\theta$. From elementary geometry the lines of flow are confocal hyperbolae with A and A' as foci. Notice that the difference of distances from P to the foci $|AP| - |A'P|$ is constant along a flow line from the property of hyperbolae. As a second example, Fig. 7.2 shows a semiinfinite strip. It is easily seen that the flow lines are confocal parabolas, with the focus at the end A of the strip. A final example is a wedge: Figure 7.3 shows an example where the wedge angle is 60°. In regions where only one face of the wedge is seen the pattern is the same as for the semiinfinite strip of Fig. 7.2. From any point where both faces are seen the effect is of a uniform Lambertian source subtending an angle of 60°. Thus $|\mathbf{J}|$ is constant in this region and the lines of flow are parallel as indicated.

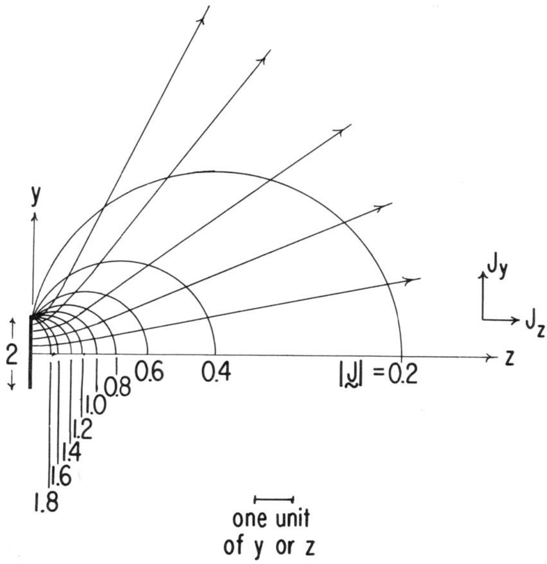

Fig. 7.1 Lines of flow and loci of constant $|\mathbf{J}|$ for the two-dimensional strip: the width is taken as two units. The axes of circles are labeled with the values of $|\mathbf{J}|$. The two sets of lines are not orthogonal, because $\mathbf{J}$ is not derived from a scalar potential.

Figure 7.4 shows a truncated 60° wedge $RQQ'R'$ which is infinitely extended back from the apex with an included angle $2\theta_o = 60°$. We have to consider four different regions in calculating vector flux fields; these are separated by the boundary lines of the wedge produced, as shown in broken lines, and they are labeled regions A (and A'), B, C (and C'), and D. A point P in region A receives flux only from the face RQ, so that the $\mathbf{J}$ distribution is the same as for a semiinfinite strip, i.e., the flow lines are confocal parabolas with focus at Q and the loci of constant $|\mathbf{J}|$ are straight lines radiating from Q. If P is in region B, it receives flux only from the strip QQ' and again we have the result that the lines of flow are confocal hyperbolas with Q and Q' as foci and the loci of constant $|\mathbf{J}|$ are arcs of circles through QQ and Q'. If P is anywhere in region D, the source has the constant angular subtense $2\theta_o$; thus $|\mathbf{J}|$ is constant at $2 \sin \theta_o$ throughout this region and the lines of flow are straight lines perpendicular to QQ'. Finally, in region C the point P receives radiation from RQ and QQ' and the angular subtense is 2θ as show in the diagram. The direction

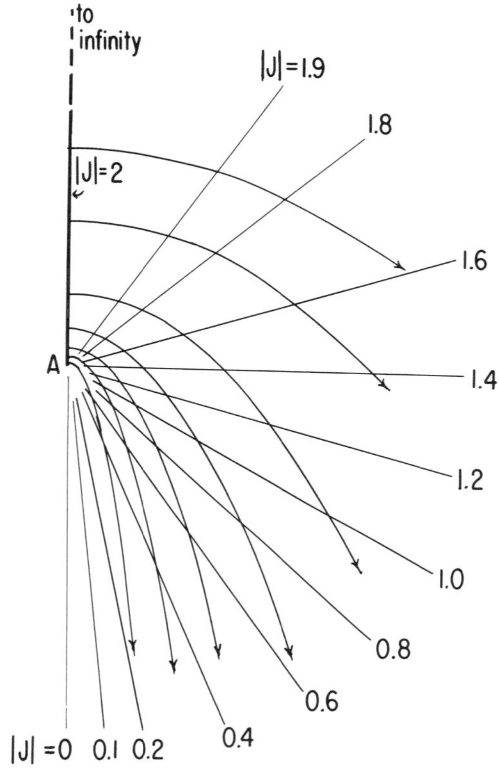

Fig. 7.2 Lines of flow and loci of constant $|\mathbf{J}|$ for a semiinfinite strip.

of **J** thus bisects the angle between a constant direction (parallel to QR) and the radius from a fixed point Q'; thus the lines of flow in region C are confocal parabolas with focus at Q' and axis through Q' and parallel to RQ. The loci of constant $|\mathbf{J}|$ are straight lines radiating from Q', since the angle 2θ as constructed is constant along these lines. The lines of flow for a truncated wedge are shown in Fig. 7.5.

7.3 3D Example

A disk as a Lambertian radiator, e.g., the circular opening of a black body cavity, is the most important example to consider. The com-

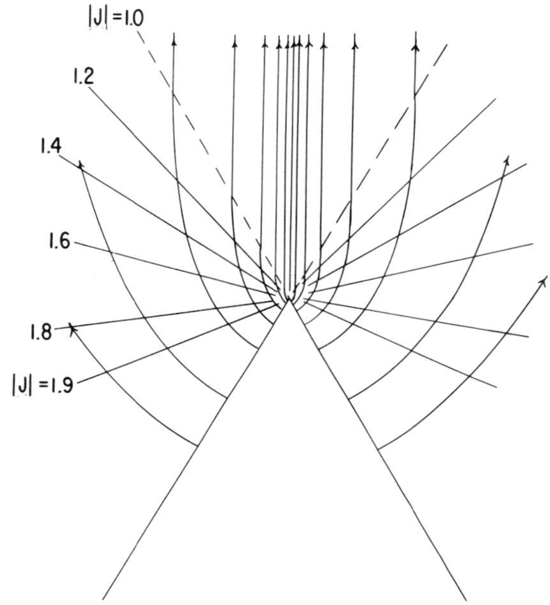

Fig. 7.3 Lines of flow and loci of constant |**J**| for a wedge.

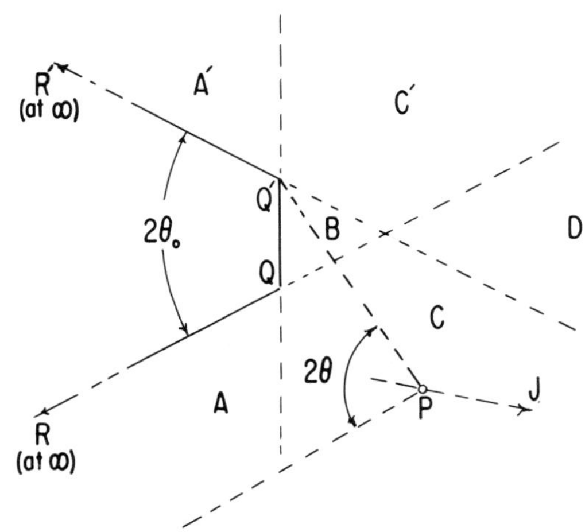

Fig. 7.4 A truncated wedge of angle $2\theta_o$. The truncated QQ' is shown as forming equal angles with the two sides of the wedge, but in fact the argument is still valid if this is not so.

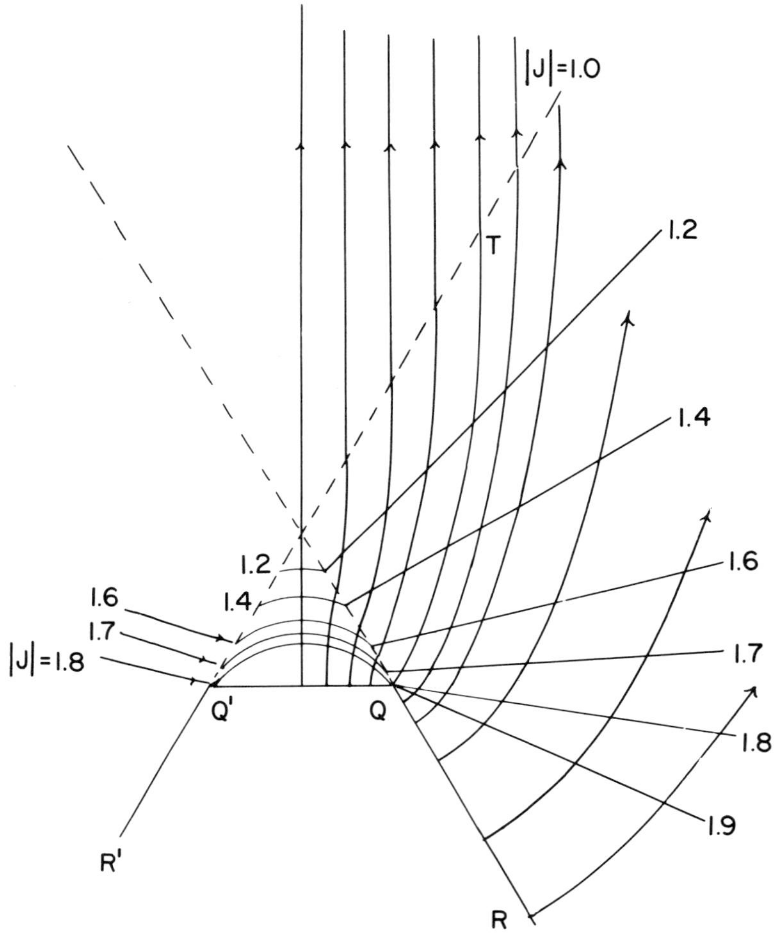

Fig. 7.5 Lines of flow of **J** and labeled loci of constant |**J**| for the truncated wedge; only the loci below the line of symmetry are shown. The wedge angle is 60°.

ponents of **J** can be calculated by integrating expressions such as Eq. (7.2) over the surface of the disk. These integrations have been carried out for radiative transfer purposes and we shall simply quote the results. Sparrow and Cess (1978), for example, give the details, and they also give a useful transformation of the surface integral over the disk into a line integral around its edge; this considerably simplifies the calculation.

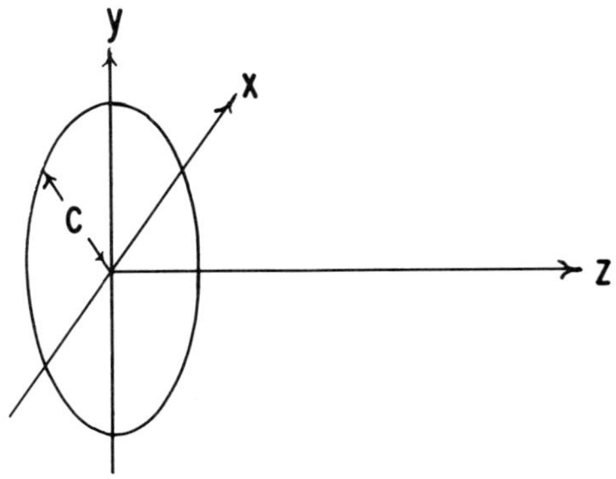

Fig. 7.6 Geometry for calculating **J** for a luminous disk.

Let c be the radius of the disk and take the coordinate axes as in Fig. 7.6. Then the components of **J** in the y, z plane, which from symmetry are all that need be considered, are given by

$$\mathbf{J}_y = \frac{\pi}{2}\frac{z}{y}\left(\frac{c^2+y^2+z^2}{[(c^2+y^2+z^2)^2-4c^2y^2]^{1/2}}-1\right)$$
$$\mathbf{J}_z = \frac{\pi}{2}\left(\frac{c^2-y^2-z^2}{[(c^2+y^2+z^2)^2-4c^2y^2]^{1/2}}+1\right) \tag{7.3}$$

The slope of the lines of flow is given by arc tan $\mathbf{J}_y/\mathbf{J}_z$, and when this is calculated it turns out that the lines of flow are the same hyperbolas as for the 2D case. The lines of flow in three dimensions are obtained by rotation about the z axis as in Fig. 7.7.

7.4 A Simplified Method for Calculating Lines of Flow

In certain cases it is easier to calculate the étendue than to perform the integrals indicated above. In case the étendue is known, we can use the solenoidal property of **J** to find the lines of flow. For example, in the case described by Fig. 7.8 the étendue H from F, F' to A, A' is equal to $[A'F]-[AF']$ in the 2D case and $\pi^2\{[AF]-[AF']\}^2/4$ in the 3D case (the Hottel strings!). Then holding $H=\text{constant}$ means we intercept a constant flux of **J** and therefore move along a line of

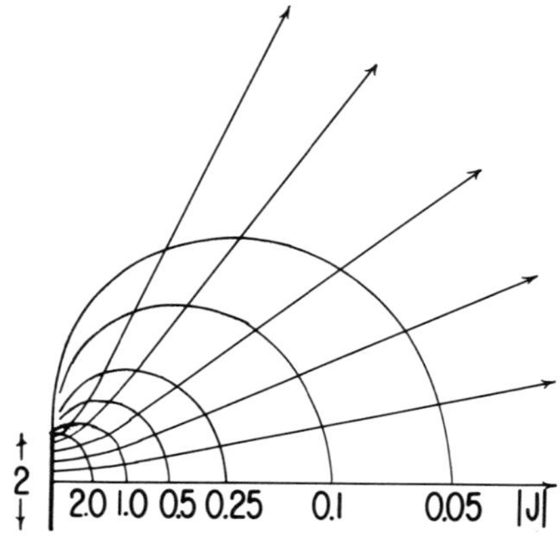

Fig. 7.7 Lines of flow and loci of constant $|\mathbf{J}|$ for a luminous disk.

flow. But holding the difference of distances to two foci, F, F' constant is the geometric definition of a hyperbola. Hence we recover the hyperbolic flow lines with lighter computation. By similar reasoning we can also obtain the components of $\mathbf{J}$. Thus the relations $\mathbf{J}_z = (\partial H/\partial y)/(2\pi y)$ and $\mathbf{J}_y = -(\partial H/\partial z)/(2\pi y)$ give Eq. (7.3).

7.5 Properties of the Lines of Flow

Let $\delta\Sigma$ be a small surface element with a 100% reflecting mirror surface on one side. If we place $\delta\Sigma$ in a vector flux distribution at a

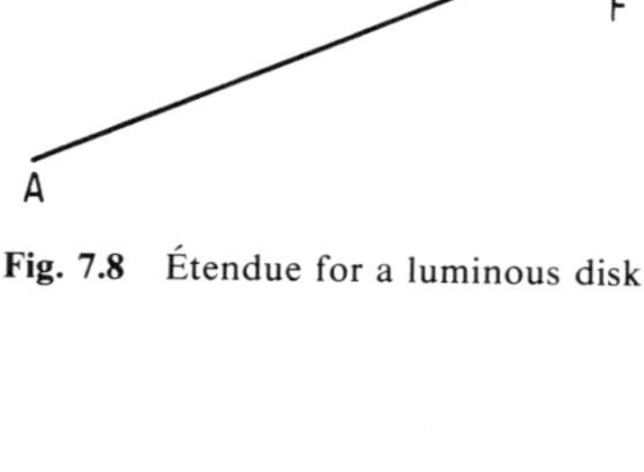

Fig. 7.8 Étendue for a luminous disk.

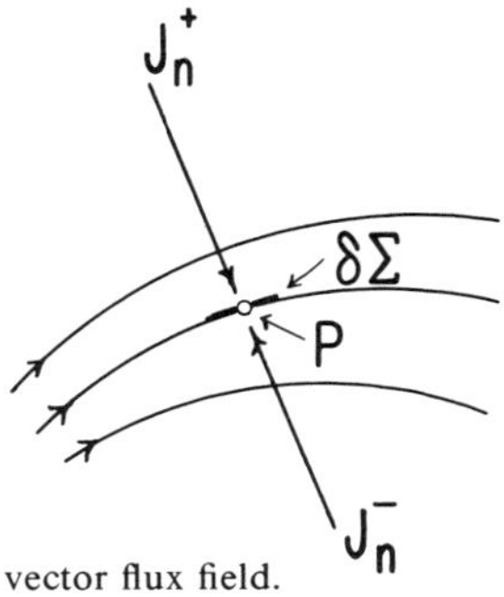

Fig. 7.9 Placing a mirror in a vector flux field.

point P (Fig. 7.9) so the lines of flow lie in its surface, then the lines of flow locally on the mirror side will be undisturbed. To see this, we note that before the mirror is placed at position P, the resolved part of **J** normal to the surface is zero, since **J** lies along the lines of flow. Now this zero normal component is made up of equal and opposite components $\mathbf{J}_n^+$ and $\mathbf{J}_n^- = -\mathbf{J}_n^+$ from rays coming from either side of $\delta\Sigma$. But the effect of inserting the mirror is to remove $\mathbf{J}_n^-$ and replace it by $\mathbf{J}_n^+$ with reversed sign, according to the law of reflection, so there is no local disturbance.

The insertion of a mirror can, of course, disturb the lines of flow at some distance, because it does not follow from $\mathbf{J}_n^- = -\mathbf{J}_n^+$ that each ray from one side is paired with a ray in the same plane of incidence and with the same angle of incidence on the other side. However, this simple situation does hold for a 2D vector flux field illuminated by a Lambertian radiator without obstructions and in the important case of a 3D field produced by a Lambertian disk.

7.6 Application to Concentrator Design

We have noted that a mirror surface placed along lines of flow of the vector flux will not disturb the overall flow pattern under a certain condition. This condition is that the zero normal component of **J** must result from the fact that, for each ray r_1 incident on one side of the surface, there must be a corresponding ray r_2 incident on the other side such that r_2 is in the direction of the reflection of r_1; we may call this a condition of detailed balance of the rays. It is possible to have $\mathbf{J}_{\text{normal}} = 0$ at a surface containing the lines of flow without the condition of detailed balance being fulfilled, and then **J** would remain undisturbed locally but not over the entire illuminated region. Clearly, the condition of detailed balance holds for a 2D field illuminated by a lambertian source without obstructions.

We may therefore place mirrors along any of the flow lines of Fig. 7.5 without disturbing the flux patterns. For example, suppose we place a mirror along the flow line from Q, ending at the point T where the flow line meets the continuation of the wedge boundary $R'Q'$, and similarly another mirror $Q'T'$ symmetrically on the other side. These mirrors take the Lambertian flux ($|\mathbf{J}| = 2$) from QQ' and convert it to a uniform flux with $|\mathbf{J}| = 1$ at TT'. The rays in the emergent beam must spread over an angle $\pm 30°$, since $|\mathbf{J}| = 1$. Clearly, we have constructed a concentrator in reverse and our construction with parabolas with tilted axes is precisely that for the compound parabolic concentrator.

Since light rays, and therefore flow lines of **J**, are reversible, we see that a way to construct a concentrator with maximum theoretical concentration is to place mirrors in the flow lines under conditions of detailed balance of the rays. This constitutes an entirely new perspective on concentrator design.

7.7 The Hyperboloid of Revolution as a Concentrator

The results of the previous section lead to another interesting new perspective on the design of nonimaging concentrators. Suppose we start with the disk radiator and we place a mirror in the form of a hyperboloid of revolution coincident with a set of flow lines as in Fig. 7.10. We truncate the mirror at some distance so the open end is a circle of diameter AA'. Considering the inside as a mirror, this forms a nonimaging concentrator with unusual properties. The foci of the hyperbolas in the section shown are at F and F', the ends of the diameter of the original disk. Then all rays entering the aperture AA' and pointing somewhere inside the disk will be reflected by the mirror so as to strike, eventually, the inner disk of diameter BB'. Thus the concentrator takes all rays from the *virtual* source FF' which can pass the entry aperture AA' and concentrates them into an exit aperture BB'.

This result is easily proved for rays in a meridional section, i.e., in the plane of the diagram. Consider a ray passing through A' and aimed at F, as indicated by the double arrow; after the first reflection it will,

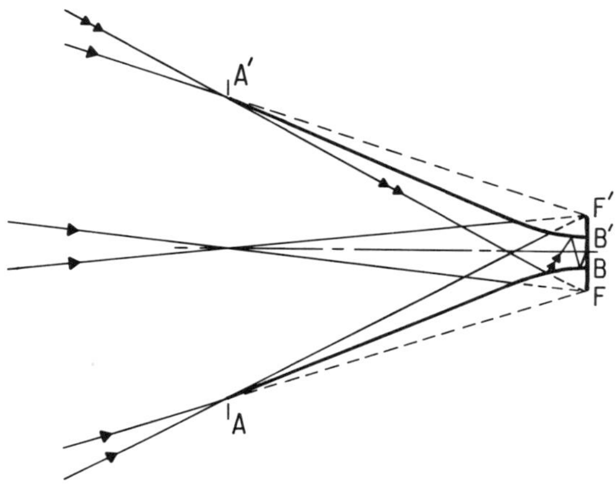

Fig. 7.10 The hyperboloid of revolution as a concentrator.

by a basic property of conic sections, be aimed at F, and so on as indicated. Thus the extreme angle rays emerge from the exit aperture, but only after an infinite number of reflections. It is easily seen that rays at angles inside the extreme angles all emerge. Thus in the meridional plane this is a concentrator of maximum theoretical concentration. This property also holds for skew rays, although this is not quite so obvious. We give in Appendix J proof that any skew ray incident on the mirror and pointing on the rim of the disk FF' will be reflected to another point on the rim and the result follows directly. When used in reverse, the same design produces a virtual ring which fills the space between a Lambertian source of diameter BB' and the larger diameter FF'. The visual effect produced is striking.

7.8 Elaborations of the Hyperboloid: The Truncated Hyperboloid

The properties of the truncated hyperboloid are evident from the above discussion. We refer to Fig. 7.11, where the hyperboloid stops short of the plane of the virtual source. For example, rays collected from the source of diameter FF' by the small aperture will appear to originate from the *same* source upon emerging from the large aperture. Alternatively, rays directed toward the source of diameter FF' at the

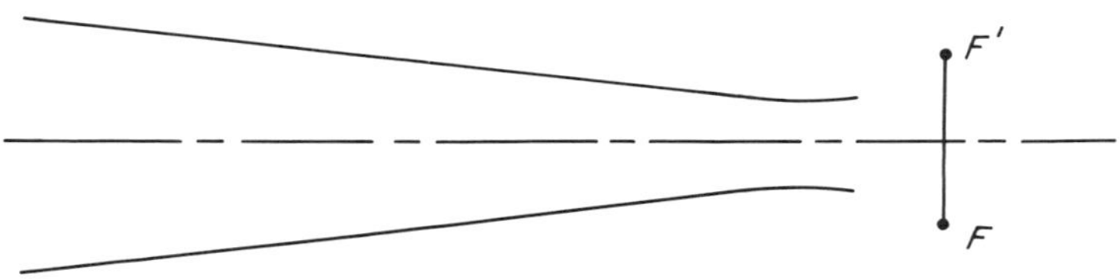

Fig. 7.11 The truncated hyperboloid of revolution as a concentrator.

large aperture will illuminate the *same* source upon exiting the small aperture. The truncated hyperboloid may be useful on its own, but usually in combination with lenses, as discussed below.

7.9 The Hyperboloid Combined with a Lens

We have seen that, for the bare hyperboloid, both real and virtual Lambertian sources are in the same plane. This is a limitation in certain applications; however, we can manipulate these positions with lenses to place the sources at more convenient locations. We show in Fig. 7.12 the full hyperboloid with a positive lens of diameter $AA' = 2a$ at the large aperture, which takes the source plane to infinity, i.e., the source is in the focal plane of the lens. Consider the properties of this design. For simplicity we neglect off-axis aberrations of the lens; a complete discussion is found in Welford, O'Gallagher and Winston

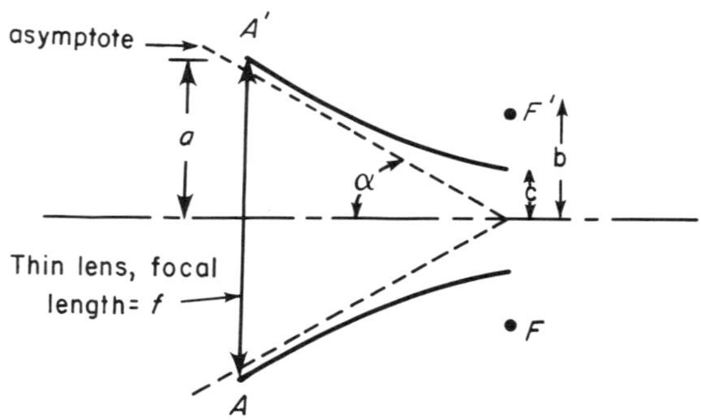

Fig. 7.12 The lens/flow-line concentrator. The entry aperture is FF', and the exit aperture is at the waist of the hyperboloid. The semiangle of the asymptotic cone is arc sin c/b, and $(a/c)^2$ is the concentration ratio.

(1987). We only remark that, since the requirements on the lens need not be severe (e.g., the focal ratio may be chosen large, as will be shown below), the aberrations can be made negligible in practice. In this approximation, the cone of rays of angle θ is imaged by the lens onto the circle of diameter $FF' = 2b$. The hyperboloid collects the rays that fill this cone and concentrates them to a smaller circle of diameter $BB' = 2c$. We show in Appendix K that $c/a = \sin\theta$. Therefore this is an ideal concentrator. This result could have been anticipated from the fact that we have postulated an ideal lens while the hyperboloid on its own is an ideal concentrator, albeit operating on a virtual source.

The design considerations readily follow from the geometry of a hyperbola. With reference to Fig. 7.12, the focal ratio of the lens can be estimated from the slope α of the asymptote to the hyperbola. The slope is given by $\sin\alpha = c/b$; the acceptance angle θ determines the focal length of the lens, completing the design.

7.10 The Hyperboloid Combined with Two Lenses

We have seen that we can construct a 3D concentrator approaching ideal properties by placing a suitable lens at the large aperture of the hyperboloid. Similarly, we can construct a 3D θ_1/θ_2 concentrator (angle transformer) by placing an additional negative lens at the small aperture of the truncated hyperboloid, which moves the source plane to infinity as shown by Ning (1988). The construction is shown in Fig. 7.13. It is apparent from the figure that the focal lengths of the two lenses are related by $F_1/F_2 = \tan\theta_1/\tan\theta_2$ while their focal ratios are related as $f_1/f_2 = \cos\theta_1/\cos\theta_2$. We remark that this design lends itself to all dielectric versions, since the positive and negative lenses are readily molded into the shape of the transformer while in practical cases the condition for total internal reflection is likely to be satisfied.

7.11 Generalized Flowline Concentrators with Refractive Components

In 2D it is possible to generalize the flow line construction following the notions expressed in Section 7.4. For simplicity, we confine our discussion to systems which are symmetrical transverse to the optic

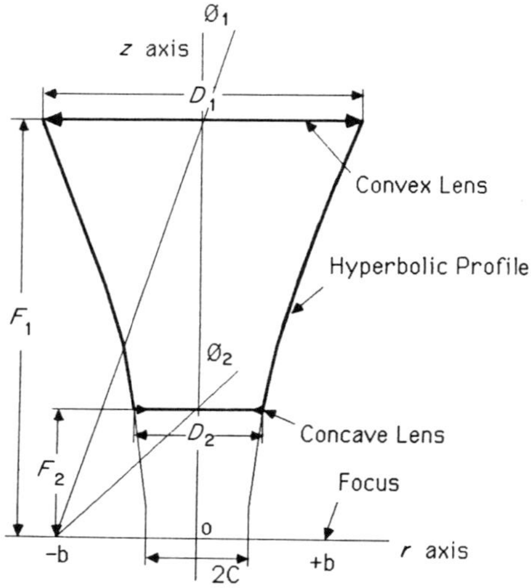

Fig. 7.13 $\theta_i \neq \theta_o$ transformer using an FLC with two lenses.

axis. As already noted in Section 7.4 the étendue H generalizes to the difference of optical path lengths (up to an overall constant). This remains true even in the presence of refractive media, provided the optical path lengths are measured along rays. These rays need not be straight lines. Thus in Fig. 7.14, the étendue H from Lambertian source

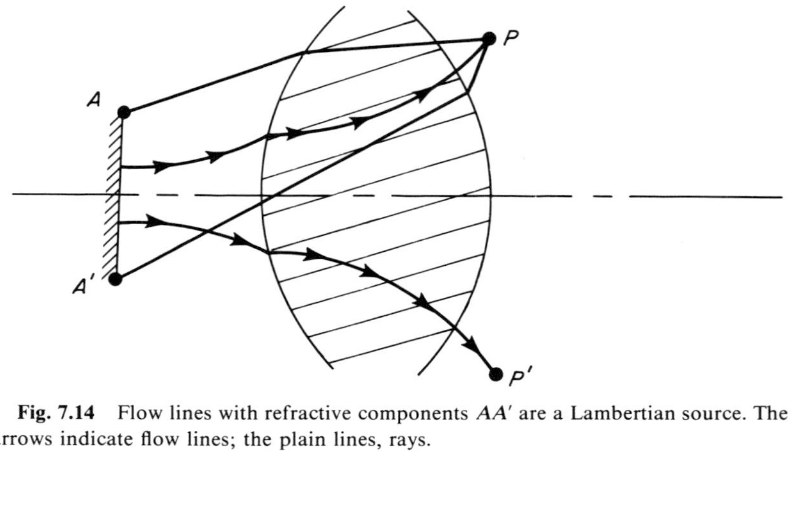

Fig. 7.14 Flow lines with refractive components AA' are a Lambertian source. The arrows indicate flow lines; the plain lines, rays.

AA' to section PP' is proportional to $[A'P]-[AP]$ where the brackets indicate optical path lengths. It follows that the lines of flow (indicated by arrows in the figure) lie along contours of $H=$ constant. Since the detailed balance condition holds in 2D, we may construct concentrators by placing mirrors along the flow lines. However, it does not follow that the 3D construction obtained by rotating the 2D flow line about the optic axis will automatically satisfy detailed balance. Specific cases will have to be checked with respect to detailed balance before the usefulness of the 3D designs can be evaluated.

CHAPTER 8

Physical Optics Aspects of Concentrators and Collectors

8.1 Introduction

The geometrical optics model of nonimaging optics is well defined in terms of rays, Fermat's principle, and Liouville's theorem, leading to the concept of phase space volume or étendue. This model is probably adequate for all large-scale applications, where by large-scale we mean systems with very large dimensions compared to the wavelength of the radiation to be handled. There are instances in which the systems are small-scale—for example, concentrators used for millimetric waves with very small detectors, and fibre optics connectors, launchers, and terminators; also the question of taking physical optics into account is, as with any optical system, of general interest. We attempt in this chapter to introduce physical optics into some of the systems described in previous chapters.

We first have to define the particular physical optics model to be used; leaving out fine distinctions, there are three levels: quantum optics, electromagnetic wave theory, and scalar wave theory. We shall refer to all three but we use mainly scalar wave theory; in this the radiation is represented by the complex amplitude, an ill-defined concept roughly equivalent to electric field strength in electromagnetic theory but with polarization ignored.

We can state the broad conclusions as follows. In most large-scale systems the effect of taking diffraction into account is that the upper bound on concentration dictated by Liouville's theorem still holds but diffractive effects mean that this upper bound can never be reached, even with no absorption, etc., losses. For small-scale systems the same conclusion generally holds, but for both large- and small-scale systems there may be resonance effects in special cases which might be regarded as invalidating this conclusion. An example from classical optics is the Fabry–Perot interferometer: in this instrument when the wavelength and angle of incidence of the incident light are correct for maximum transmission the power density inside the interferometer is equal to that outside multiplied by the finesse, a number in the range 10–100 or more. However, the angle and wavelength selectivity of the interferometer are such that it is questionable whether this constitutes a real case of concentration beyond the Liouville theorem limit, particularly as it is not possible to tap off or otherwise get at this power density without destroying the finesse and so lowering the concentration.

8.2 Étendue in the Physical Optics Model

It is first necessary to agree on what is meant by étendue or phase space volume in physical optics. We recall that the discussion in geometrical optics starts with the quantity

$$dH = n^2\, dx\, dy\, dL\, dM \tag{8.1}$$

which is a differential element of phase space or étendue. This was shown to be conserved along a ray apart from reflection, absorption, or scattering losses. Now recall the definition of radiance in geometrical optics as flux per unit solid angle per unit area projected perpendicular to the ray direction; then apart from dimensional and scaling factors it can be seen that dH is a differential element of radiance from a Lambertian source in the x, y plane. Thus étendue is radiance integrated over area and direction cosine space, or radiance, is a differential element of étendue, and it is conserved along a ray with the same provisos as above.

One way to make progress with a physical optics model would, therefore, be to get a physical optics version of radiance; this turns out to be very difficult, for a fundamental reason—that power and energy flow in a wave field can only be uniquely defined as averages over finite volumes, whereas radiance in our sense is concerned with energy flow at a point and in a certain direction. For example, it has been shown that in the detailed structure of the monochromatic point spread function near a focus there are some regions where the energy flow is in the reverse direction to the general direction of travel of the light (Stamnes, 1986); it would be difficult to define radiance in a direction passing through such a region because for other positions along this direction the flow would be in the opposite direction. The first attempt to get a definition of radiance in the physical optics domain was made by Walther (1968) and the same author offered some alternatives (Walther, 1973). Several authors then began to study this topic and publications continue to appear at the time of writing. It became apparent that there could be several definitions of a scalar function of position and direction $B(\mathbf{r}, \mathbf{s})$ such that, when integrated over a volume, the correct result for power flow was obtained but none fulfilled all the obvious physical requirements such as being real and nonnegative and propagating approximately in a straight line, except under restricted conditions on the radiation source. Wolf (1978) discussed a theorem due to Friberg which proves that no radiance function can satisfy simultaneously all four of the following conditions: (1) it should be a linear function of the cross-spectral density, (2) it should be nonnegative, (3) it should be zero for directions not originating in a source, and (4) when integrated over area it should give the radiant intensity. It turns out also that when such *generalized radiance functions* are used to calculate optical images, the results do not agree with ordinary image formation calculations. However, these functions are the best thing available if we wish to carry out phase space-type calculations in physical optics, and in Section 8.4 we give an example of the use of one of them.

8.3 Defining Generalized Radiance

An immediate complication in attempting to define generalized radiance is that the radiation field may be partially coherent, that is, it may be possible to observe interference effects between light beams sampled from different parts of the field, whereas in the geometrical optics model this is by definition impossible: All rays intersect without interference and they meet obstacles without diffraction. Likewise in geometrical optics light does not have a wavelength, whereas for the present purposes of physical optics the radiation has a finite but small spread of wavelengths; it is said to be quasimonochromatic.

Bearing the above in mind we consider a radiation field with position vector $\mathbf{r}$ and complex amplitude $u(\mathbf{r})$. Then the generalized radiance according to Walther (1973) is (see Appendix I)

$$B(\mathbf{r}, \mathbf{n}) = \mathrm{Re}\langle u^*(\mathbf{r})\bar{u}(\mathbf{n}) \exp(ik\mathbf{n} \cdot \mathbf{r})\rangle \tag{8.2}$$

where $\bar{u}(\mathbf{n})$ is defined as in Appendix I. In Eq. (8.2) the brackets $\langle\ \rangle$ indicate the ensemble average over waves from all points in the original source; this takes account of the partial coherence due to the finite source size.

The expression in Eq. (8.2) is only one of many similar expressions which have been suggested in an extensive literature; as mentioned in Section 8.2, none of these fulfills the conditions which radiance ought to satisfy, e.g., it ought to be always real and positive or zero, and it should be zero in directions which do not originate at a source. However, they do conform to the requirement that the radiance integrated over a closed surface gives the net power balance. It will be seen in later parts of this chapter that generalized radiance is of limited utility in practical problems; nevertheless, we shall use it in one attempt to set a limit on concentration using the scalar wave model.

8.4 Efficiency of 2D Concentrators in the Scalar Wave Model

In this section we outline the proof due to Winston and Welford (1982b) that the geometrical optics limit set by Liouville's theorem is an upper bound in scalar wave theory. Figure 8.1 shows a 2D concentrator with an incoming field $u(x)$ at $z = 0$; we write this field in terms of its Fourier transform as

$$u(x) = \int \bar{u}(L) \exp(ikLx)\, dL \tag{8.3}$$

and from Eq. (8.2) we take the radiance to be

$$B(x, L) = \mathrm{Re}\langle u^*(x)\bar{u}(L) N \exp(ikLx)\rangle \tag{8.4}$$

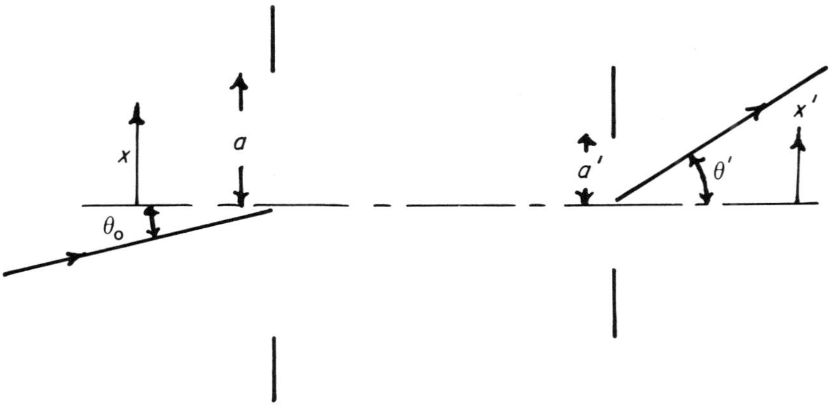

Fig. 8.1 Notation for a concentrator.

The entry étendue is

$$H = 4Ba \sin \theta_0 \tag{8.5}$$

and the Fourier transform of the mutual intensity is

$$\Gamma(L, L') = \langle \bar{u}(L')\bar{u}^*(L)\rangle = \{2aB/\lambda N\} \operatorname{sinc} ka(L' - L) \tag{8.6}$$

Now according to Gabor (1964), a beam of finite width $2a$ can carry only a finite amount of information and it can therefore be expressed in terms of a finite number, say, f, of independent functions, where $f = 4(a/\lambda) \sin \theta_0$. Following Walther (1967), let these be a set of orthonormal functions $\phi_n(L)$ at $z = 0$, so that

$$\int_{-L_1}^{L_1} \phi_m^*(L)\phi_n(L) N\, dL = \delta_{mn} \tag{8.7}$$

Expanding Γ, we have

$$\Gamma(L, L') = \sum_{mn} A_{mn}\phi_m(L')\phi_n^*(L) \tag{8.8}$$

with

$$A_{mn} = \int \phi_m^*(L')\phi_n(L) N' N \Gamma(L', L)\, dL'\, dL \tag{8.9}$$

Then from Eqs. (8.6) and (8.7) and remembering that the ϕ_m and sinc are bandwidth limited we find

$$A_{mn} = B\delta_{mn} \tag{8.10}$$

Thus the correlation matrix is diagonal and the net entry flux is $\lambda \operatorname{Tr} A_{mn} = \lambda f B = 4aB \sin \theta_0$ as in Eq. (8.5).

To propagate from the entry aperture to the exit aperture, we assume only a linear transformation of complex amplitudes,

$$\bar{u}'(L') = \int \bar{u}(L) K(L, L')\, dL \tag{8.11}$$

To transform intensities, we use the hermitian kernel

$$E(L, L_0) = \int K(L, L') K^*(L_0, L') N'\, dL' \tag{8.12}$$

and we have

$$\int \phi_i^*(L_0)\phi_i(L) E(L, L_0)\, dL\, dL_0 = \int \tau_i \phi_i^*(L)\phi_i(L) N\, dL \tag{8.13}$$

In this expression the τ_i are transmission factors for the orthonormal fluxes and they have to be such that $0 < \tau_i \leq 1$. The emerging ψ_i from each ϕ_i is given by

$$\psi_i(L') = \int K(L, L')\phi_i(L)\, dL \tag{8.14}$$

so that

$$\int \psi_i(L')\psi_j^*(L') N'\, dL' = \int E(L, L_0)\phi_i(L)\phi_j^*(L_0)\, dL\, dL_0 \tag{8.15}$$

But we can specify the ϕ_i to be solutions of the integral equation

$$\int E(L, L_0)\phi_i(L)\, dL = \tau_i \phi_i(L_0) N_0 \tag{8.16}$$

so that the right-hand side of Eq. (8.15) becomes

$$\int \tau_i \phi_i(L_0)\phi_j^*(L_0) N\, dL_0 = \tau_i \delta_{ij} \tag{8.17}$$

Then the output for $\Gamma(L', L)$ is

$$B \sum_i \psi_i(L')\psi_i^*(L) \tag{8.18}$$

and the total exit flux is

$$\lambda B \Sigma \int |\psi_i(L')|^2 N' \, dL' = \lambda B \Sigma \tau_i \tag{8.19}$$

Thus the exit flux is less than or equal to $\lambda f B$, as was to be shown.

If the angular extent of the output is $\pm\theta_0'$ over width $2a'$ the number of degrees of freedom in the output will be $f' = (4a'/\lambda)\sin\theta_0'$, and if this is equal to f the exit geometrical étendue will be equal to that at the entry, but the physical exit étendue must be equal to or less than the physical entry étendue.

Thus physical optics étendue is either conserved or decreased on transmission through an optical system. This treatment is extended to 3D systems by Winston and Welford (1982b).

8.5 Efficiency of 3D Concentrators: An Image-Formation Approach

The treatment in Section 8.4 is open to criticism on account of its use of the generalized radiance, a concept to be used with great caution. In this section we take an entirely different approach to the problem of setting limits to concentration in physical optics. The method is that of Welford and Winston (1982a).

The difficulty in dealing with nonimaging systems by scalar wave theory, or indeed by any optical model except geometrical optics, is that there is in, e.g., a CPC, an infinite number of possible paths involving numbers of reflections from zero to infinity. Thus, in imaging optics terms, there is an infinite number of highly aberrated point spread functions to handle at the exit aperture; of course, only a finite number of these carry an appreciable fraction of the total étendue but the number is great enough to make a straight image-formation approach impracticable. We therefore adopt the postulate that the best possible nonimaging concentrator cannot exceed in efficiency the performance of an ideal imaging optical system used as a concentrator, and we consider instead the limits on performance of such an imaging system. This may seem an unfruitful approach, since it is accepted that real optical systems cannot be designed to better than half the concentration ratio of a nonimaging system in 2D or one quarter of a system in 3D. Nevertheless, there is nothing to stop us from considering an ideal imaging system as a concentrator.

Figure 8.2 shows a 2D imaging system used as a θ_i/θ_0 concentrator with entry and exit angles β and α, respectively. Let the entry and exit apertures be, respectively, $2a$ and $2h$. The maximum possible geometrical concentration ratio is, by Liouville's theorem, $\sin\alpha/\sin\beta$ and the concentrator will achieve this if

$$h = a \sin\beta/\sin\alpha \tag{8.20}$$

To fix our ideas, we could imagine β to be 1° and α, 30°; with small linear dimensions this would correspond to a medium-power microscope objective and these can be designed to be aberration free over a small field and small wavelength range.

We can use well-known image-formation methods to calculate in 2D the light intensity distribution at the plane of the exit aperture as a function of position h, due to a uniform source at infinity subtending $\pm\beta_0$ and producing light power W inside $\pm\beta_0$. It is

$$(W/2\alpha)\sin\alpha\{G(z_0+z)+\operatorname{sgn}(z_0-z)G(|z_0-z|)\} \tag{8.21}$$

where

$$G(x) = (2/\pi)\{\mathrm{Si}(2x) - x^{-1}\sin^2 x\} \tag{8.22}$$

and $z = kh\sin\alpha$, $z_0 = ka\sin\beta_0$. Only positive values of z and z_0 are

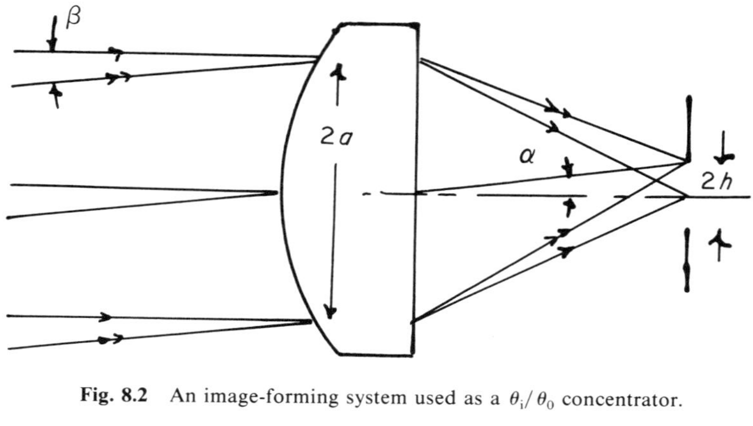

Fig. 8.2 An image-forming system used as a θ_i/θ_0 concentrator.

to be taken, as the distribution is symmetrical about $h=0$. For x greater than 7 one can take

$$G(x) \sim 1-(1/\pi)\{x^{-1}+\tfrac{1}{2}x^{-2}\sin 2x\} \tag{8.23}$$

to four decimal places.

The coefficient of $(W/a)\sin\alpha$ in Eq. (8.21) is the normalized image of an object subtending $\pm\beta_0$ and this can be compared with a rectangle function of unit height and width $\pm z_0$; the ratio of the areas under the two within $\pm z_0$ gives the efficiency of the system as a concentrator. Obviously, some light always spills outside the geometrical optics exit aperture, so the system is never perfect. Figure 8.3 shows some of the normalized images from Eq. (8.21) and Fig. 8.4 shows the efficiency as a function of z_0.

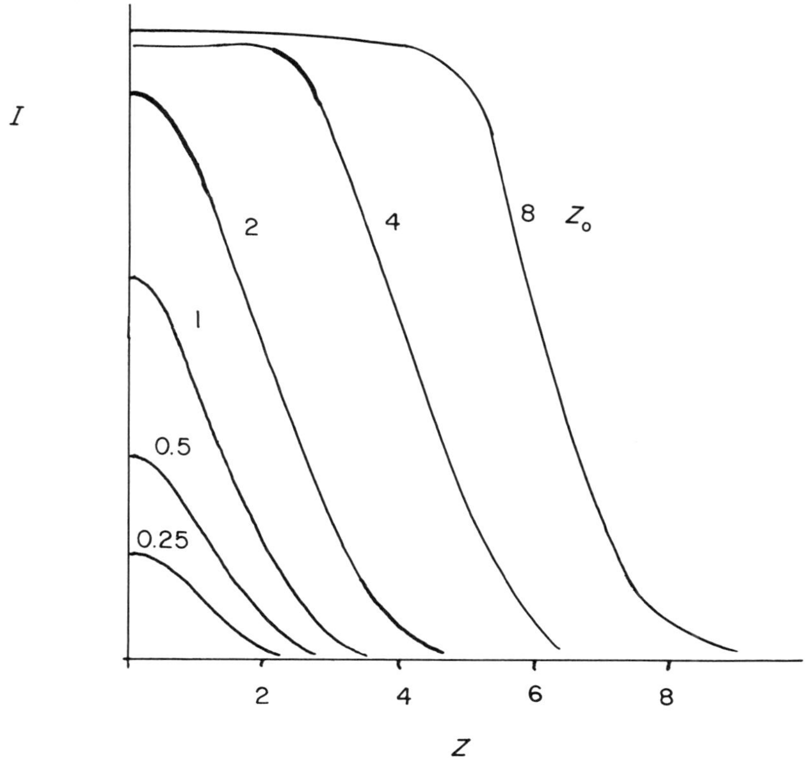

Fig. 8.3 Images from Eq. (8.21).

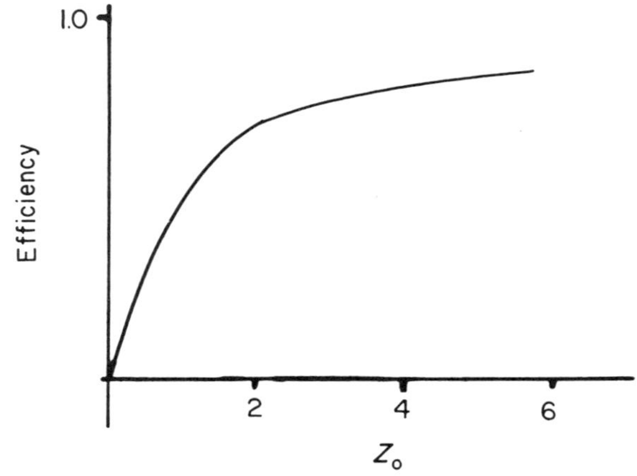

Fig. 8.4 Efficiency as a function of scaled aperture z_0.

Of course, the curves in the figures all apply to exit apertures only a few spread function widths in size, since the width at the first zeroes of the 2D spread function is $\pm\pi$, but they indicate clearly that in physical optics, if the system has the ideal geometrical concentration ratio, it cannot reach 100% efficiency.

A similar calculation can be done in 3D and it was done by Hildebrand and Winston (1982). The point spread function then involves a Bessel function instead of a sine and the numerical work is that much more laborious, but the conclusions are similar to those for the 2D case. However, this work is interesting in that it was applied to a particular experiment with millimetric waves and with an entry aperture only some tens of wavelengths in diameter; the experimental results for concentration as a function of entry aperture diameter agreed well with theory.

8.6 The Quantum Optics Approach

From a quantum optics point of view a light beam meeting a concentrator is modeled as a stream of particles meeting a potential distribution and since in this picture there is always some proportion of backscatter, the general inequality that the Liouville limit of concentration cannot be reached must hold. The first detailed study of this model was made by Bassett and Winston (1984). These authors considered the propagation of *specific intensity* I $(\mathbf{r}, \mathbf{n}, \omega)$ through an optical system; specific intensity is defined as power per unit area per unit solid angle per unit angular frequency range, so that it corresponds to radiance for polychromatic light in physical optics. The analog in quantum mechanics of specific intensity is the density in phase space $\sigma(\mathbf{r}, \mathbf{p})$ for a particle in a potential distribution. Let σ_{max} be the maximum density in phase space of an incident particle and let σ' be the density anywhere after scattering. Then it was shown by Bassett and Winston (1984) that

$$\sigma' \leq \sigma_{max} \tag{8.24}$$

everywhere in the scattering region.

This result can then be applied to the scattering of photons to give the inequality for specific intensity

$$I' \leq I_{max} \tag{8.25}$$

This result can be put approximately into the terminology used in most of this book as follows: The radiance at any point and in any direction at the exit aperture of a concentrator cannot exceed the maximum radiance at the entry aperture. This is a stronger result than those of Sections 8.4 and 8.5 since it applies to specific intensity rather than to étendue and since it does not apply only to related initial and final "rays" but compares the specific intensity at *any* point and direction in the exit with the maximum in the entry.

8.7 Resonance Effects

As suggested in Section 8.1, increasing energy density by means of resonance as in a Fabry–Perot cavity is not a fair example of concentration in the context of this book, since it is not possible to extract the energy without spoiling the resonance and thus lowering the concentration. However, we note that this topic has been studied in detail by Bassett (1984, 1985) using both quantum mechanics and electromagnetic theory approaches.

We can state Bassett's general conclusion briefly as follows. Let $I_{\max}$ be the maximum incident specific intensity on a resonant structure and let R be a typical dimension of the structure; then an upper bound on the energy density in the structure is $8\pi^2 I_{\max}/R$; in more detail, this has to be expanded to define R more carefully and to discuss how closely the upper bound may be approached but for our purposes it is enough to point out that an upper bound exists, since this is not obvious in classical optics. Thus consider our original example of the Fabry–Perot cavity: Although in practice it is difficult to reach a finesse greater than about 100 even with small diameter interferometers because of coating and flatness problems, we can postulate as high a finesse as we please and then there would seem to be no limit on the attainable energy density; however, as the finesse increases, the cavity necessarily becomes more selective in angle and wavelength, and it is this which produces the upper bound on energy density.

8.8 Focusing in Electromagnetic Theory

In Section 8.5 and in earlier chapters of this book we discussed image formation as concentration using both geometrical optics and scalar wave theory. Bassett (1986) studied this problem in electromagnetic theory, considering an unspecified optical system with no resonance effects, i.e., only one transit through the system. He obtained the following simple result.

If total light power W of wave number $k(=2\pi/\lambda)$ is concentrated, then an upper bound on the attainable energy density at a focus or anywhere else (for this applies to nonimaging systems also) is $k^2 W/(3\pi c)$, where c is the speed of light.

This may be compared to an elementary result from scalar wave theory: If a lens with a circular aperture forms an aberration-free point spread function from an incident beam of total power W, then the maximum possible power density in the point spread function is $\pi W/\lambda^2$, assuming a numerical aperture 1 for the lens. If this is turned into energy density by dividing by c, it is found that the two formulae differ only by the factor 4/3, a result which tends to strengthen confidence in the power of scalar wave theory.

8.9 More about Generalized Radiance

We have made free use of generalized radiance and of its physical analog, specific intensity, in this chapter. In this final section we examine in some detail the relationship of generalized radiance to experimental optics.

How should radiance or specific intensity be measured? The principles are simple (see, e.g., Grum and Becherer, 1979). A collimator with a small detector at its focus is set up at the point at which the radiance is to be measured and with its axis pointing along the required direction, as in Fig. 8.5; the signal from the detector is then a measure of the radiance. The finite sizes of the detector and of the collimator aperture necessarily imply that the result is averaged over the area of the collimator aperture and the solid angle subtended by the detector. This has significance for the interpretation of generalized radiance, as will be seen below.

We now consider the calculation of generalized radiance in simple specific cases. We use a rearrangement of Eq. (8.2) also given by Walther (1973):

$$B(\mathbf{r}, \mathbf{n}) = \mathrm{Re} \int F(\mathbf{n}, \mathbf{n}') \exp\{-ik(\mathbf{n}-\mathbf{n}'), \mathbf{r}\}\, d\Omega' \tag{8.26}$$

In this expression $F(\mathbf{n}, \mathbf{n}')$ is the correlation between waves of the angular spectrum of plane waves in directions $\mathbf{n}$ and $\mathbf{n}'$; this is the Fourier transform of the cross-spectral density of the field. We apply this to a pair of plane waves, coherent with each other, of equal intensity and in respective directions $\mathbf{n}_1$ and $\mathbf{n}_2$. The angular correlation is then

$$\begin{aligned} F(\mathbf{n}, \mathbf{n}') = \delta(\mathbf{n}-\mathbf{n}_1)\delta(\mathbf{n}'-\mathbf{n}_2) + \delta(\mathbf{n}-\mathbf{n}_2)\delta(\mathbf{n}'-\mathbf{n}_1) \\ + \delta(\mathbf{n}-\mathbf{n}_1)\delta(\mathbf{n}'-\mathbf{n}_1) + \delta(\mathbf{n}-\mathbf{n}_2)\delta(\mathbf{n}'-\mathbf{n}_2) \end{aligned} \tag{8.27}$$

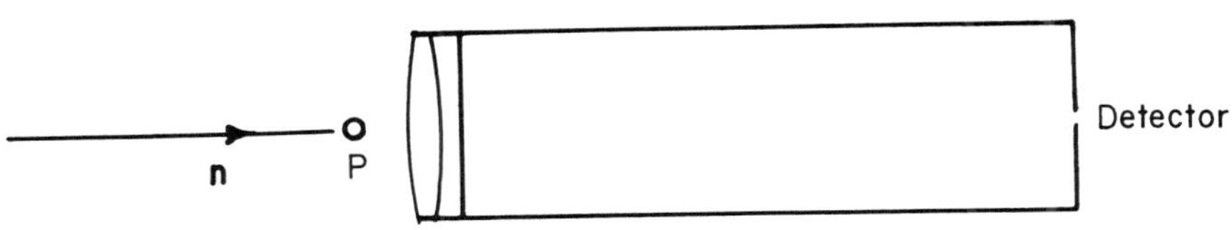

Fig. 8.5 A collimator used to measure radiance; the aperture is set at *P*, the point at which the radiance is to be measured, and it points in the direction $-\mathbf{n}$.

In this expression the first two products of delta functions express the perfect correlation between the two waves, since they are assumed to be coherent; the second two pairs state that each wave is coherent with itself. If this is substituted into Eq. (8.26), we obtain for the radiance

$$B(\mathbf{r}, \mathbf{n}) = \{\delta(\mathbf{n}-\mathbf{n}_1) + \delta(\mathbf{n}-\mathbf{n}_2)\} \cos\{k(\mathbf{n}_1-\mathbf{n}_2) \cdot \mathbf{r}\} + \delta(\mathbf{n}-\mathbf{n}_1) + \delta(\mathbf{n}-\mathbf{n}_2) \tag{8.28}$$

The two separate delta functions indicate that the radiance is nonzero only along the directions $\mathbf{n}_1$ and $\mathbf{n}_2$, which is reasonable, but the cosine term needs some explanation; it has the spatial frequency of the interference fringes formed between the two plane waves, but does it influence the measurement of radiance? If we put the radiance measuring collimator in this field, the detector gives a signal when the axis points in either of the directions $-\mathbf{n}_1$ or $-\mathbf{n}_2$, but to distinguish clearly between these directions it must have enough angular resolving power and this implies that the aperture of the collimator lens must be larger than the interference fringe spacing (Welford and Winston, 1987), so that the fringe structure is not recorded. Similar arguments apply if the plane waves are mutually partially coherent or incoherent. We conclude from this simple example that, although generalized radiance may be a useful construct for some calculations, it is not a measurable quantity and it must therefore be used with care.

Some conformity of generalized radiance with experimental optics is obtained by introducing the concept of a *quasihomogeneous source*; this was done by Carter and Wolf (1977), who showed that the generalized radiance (according to more than one different definition) from such a source is always nonnegative, that it is the product of a function of position and a function of direction, and that the function of position at the source is just the intensity. The restriction to quasihomogeneity means in simple terms that the properties of the source must vary slowly across it [the exact definition is given by Carter and Wolf (1977)] and this is too restrictive for most interesting cases. Consider, for example, a uniform Lambertian source in the form of a strip of finite width or a semiinfinite plane terminated in a sharp edge; both of these are of practical interest in applied optics but they are not quasihomogeneous. Winston and Ning (1988) used the generalized radiance expression given by Eq. (8.26) and found not only that the generalized radiance was nonzero in directions not pointing to the source region but that it was negative in certain regions.

The collimator method of *measuring* radiance described above could be applied to such calculations and we have found that, although the grosser unrealistic oscillations in the radiance curves are smoothed out, as the "fringes" would be smoothed out for the two-plane-wave experiment described above, there is still no agreement with the predictions of calculations based on ordinary image-formation theory.

8.10 Conclusions

We have to conclude, rather lamely, it must be admitted, that so far the main result from attempting to introduce physical optics into concentrator theory is the replacement of equalities by inequalities in statements about limits on concentration. This tells us that we cannot do better than the geometrical optics predictions and will in general do worse. However, for large-scale systems this is hardly relevant, since reflection losses will certainly be much greater than those from diffraction and scattering effects. For small-scale systems the result is relevant, but at present there seems little prospect of a simple theoretical approach to specific nonimaging systems.

CHAPTER 9

Shape Tolerances and Manufacturing Methods for Nonimaging Optical Components

9.1 Optical Tolerances

In the design of image-forming systems there are two kinds of optical tolerances. The first kind are theoretical tolerances for aberrations, and the second are constructional or manufacturing tolerances. We saw in Chapter 3 that it is impossible to design image-forming systems completely free from aberrations for finite aperture and field, and this gives rise to the need for aberration tolerances. Aberration in the present sense is essentially a concept of geometrical optics. The most stringent system of tolerances uses the approach of reducing the geometrical aberrations until the residual image imperfection is entirely due to diffraction effects, i.e., the performance or resolving power is substantially that which would be expected from a geometrically perfect system after allowing for the finite wavelength of the light. This system, due originally to K. Strehl, is described by, for example, Born and Wolf (1975). Other systems allow larger amounts of aberration, depending on the application for which the optical system is intended.

Constructional tolerances are concerned with the differences between the system as designed and as made, i.e., errors of surface sphericity, refractive index inhomogeneity, malcentering, etc.

9.2 Tolerances for Nonimaging Concentrators

We can immediately identify the problem of constructional tolerances for nonimaging concentrators, and most of this chapter will be concerned with the results of ray-tracing calculations done to obtain such tolerances. The property corresponding to theoretical aberration tolerances is failure of the design to have maximum theoretical concentration ratio, i.e., noncompliance with the edge-ray principle. Here the problem is not very profound since we have already seen in Chapter 4 that the basic CPC in 2D is perfect in this sense. In 3D the CPC is not quite perfect but we saw that there are no degrees of freedom available for improving the performance anyway. Similar remarks apply to FLCs. Thus there is nothing corresponding to the elaborate processes of computer optimization used in the design of imaging systems, and there is little to be said about theoretical "aberrational" tolerances.

Returning to constructional tolerances, before considering ray tracing we can propose a heuristic argument that suggests that the tolerances in the figure of, say, a basic CPC can be very large. Figure 9.1 shows a CPC with a ray at the extreme angle θ_i. We assume the shape errors on the form of the CPC to be random and such that at the point of incidence Q the reflected rays are deviated from the correct direction according to a probability density function that is symmetric about the undeviated direction. Then from the geometry of the CPC in the 2D case just half of the deviated rays will be rejected, e.g., ray QP_1', if we take an average. Also from the geometry, if the angular spread is rather small, i.e., if the probability density function is negligible outside some small angle ε, then rays at angles less than $\theta_i - \varepsilon$ will all be transmitted. Thus the effect on the angle transmission at θ_i is 50% and the theoretical square cut-off is smoothed over a

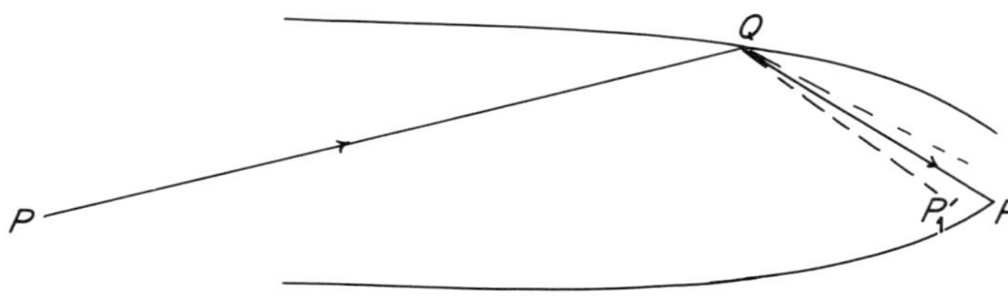

Fig. 9.1 Effect of shape errors on the efficiency of a CPC.

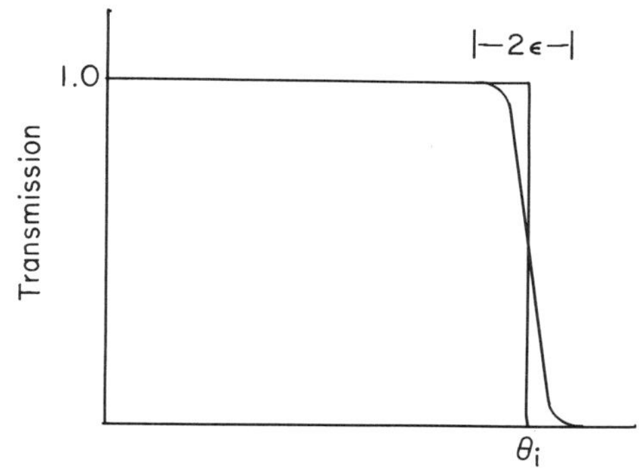

Fig. 9.2 The effect of shape errors on the transmission-angle curve of a CPC.

range $\pm\varepsilon$. The precise shape depends, of course, on the form of the slope error probability function but the curve will look as in Fig. 9.2.

The effect for a 3D CPC would be much the same, i.e., the rounded curves such as in Fig. 4.12 would become slightly more rounded.

The tolerances are thus very large. For example, in a 2D CPC designed for a concentration ratio of 10 ($\theta_i = 5.7°$) it would not be unreasonable to allow slope errors on the surface of $\pm 0.25°$, since the resultant rounding of the graph would only be over $\pm 0.5°$. In the next section we present some numerical confirmation of these ideas.

9.3 Ray-Tracing Results

We can get a reasonable grasp of the effect of slope errors by the following approximate procedure. Rays are traced through a concentrator, and at each reflection the direction of the surface normal is changed randomly by an amount $\delta\theta$ according to a Gaussian distribution of specified standard deviation (the direction of the reflected ray is deviated by $2\delta\theta$). If enough rays are traced, the result is an approximation to the ensemble average transmission-angle curve. The choice of $\delta\theta$ within the Gaussian distribution is governed by a simple algorithm based on a table of random numbers. This method ignores the fact that slope errors on the concentrator surface necessarily also imply errors in position or height of the surface, but a few sketches of typical concentrator geometry are enough to show that height variations alone cause only very slight wanderings of the rays at the exit aperture. Also, this method takes no account of the second-order statistics of the shape errors, i.e., how rapidly the slope changes from point to point (the "correlation length" for slope variations). This again is a negligible effect, since most of the rays have either one reflection or none, and very few rays are such that they would have two reflections within the correlation length if this is a small fraction of the length of the concentrator. Thus we can expect our method to give useful results for slope errors not exceeding a degree or two.

Smaller-scale irregularities of surface height have a more complicated effect and it is necessary to use rather elaborate theory to estimate their effects (see, e.g., Welford, 1977). The general effect is that roughness with lateral scale as small as a few wavelengths could diffract light quite large angles away from the correct reflection direction in a diffuse halo. Fortunately, it turns out that many surfaces suitable for manufacturing large concentrators have very small wide-angle scatter. Measurements by Pettit (1977) on materials such as aluminum-coated plastic films and rolled aluminum sheet show that more than 85% of the incident flux is reflected (95% for silver-coated films) and that the $1/e$ intensity point in the scattered light is almost always within 1° of the specular reflection direction. On this basis we can claim that our calculations are realistic.

Some results of such calculations on a 2D CPC of collection angle ±10° are shown in Fig. 9.3; the standard deviation σ_θ of the slope error is given in degrees. An approximate calculation shows that for $\sigma_\theta = 0.4°$ the transmission loss due to the shape error inside the design collection angle is about 3%, and for $\sigma_\theta = 1°$ it is still only about 8%.

It seems clear that the quality of surface shape required is relatively low compared to image-forming optical systems. In fact, as we shall see in Section 9.4 and in Chapter 10, there are sometimes actual advantages in random shape errors of moderate magnitude.

9.4 Peaks in the Emergent Light Distribution

If a basic 2D CPC collects light uniformly over its entry angle, then it follows from the theory of Chapters 2 and 4 that the exit aperture will be uniformly illuminated over a 2π solid angle. In other words, if the exit aperture is left open, the emergent light appears to come

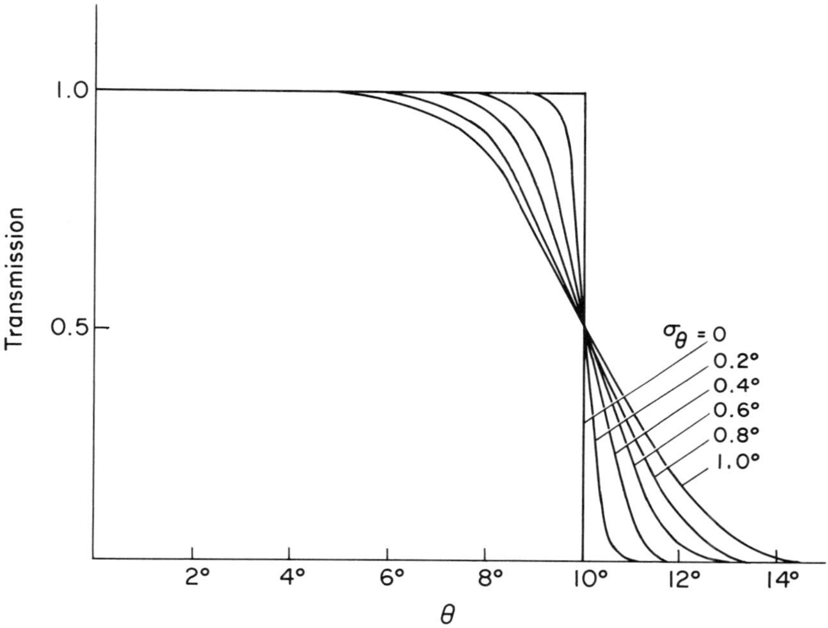

Fig. 9.3 Transmission-angle curves for 10° CPCs. The curves indicate the effect of different large-scale slope errors given by σ_θ.

from an ideal uniform Lambertian emitter. If, however, the incident light covers only a part of the entry angle, it does not follow that the emergent radiation will be uniformly distributed. In fact, there can be sharp maxima in the output. The importance of this is that in most solar energy applications the entry angle is deliberately made larger than the direct sunlight cone (±0.25°), and this, as will appear in Chapter 10, can sometimes have an adverse effect on the absorber.

In this section we shall explain the effect and show how to minimize it, since the topic is allied to that of aberrational and constructional tolerances. The applications will be explained in Chapter 10.

Let us take an extreme case first. If we have a parallel pencil incident at the extreme acceptance angle of a CPC, then all the rays emerge at one point just grazing the edge, as in Fig. 9.4, and the distribution is clearly very uneven both in angle and position. The same would hold for angles other than that shown, but owing to the aberration of the parabolic shape used as a reflector at, in effect, an off-axis angle,

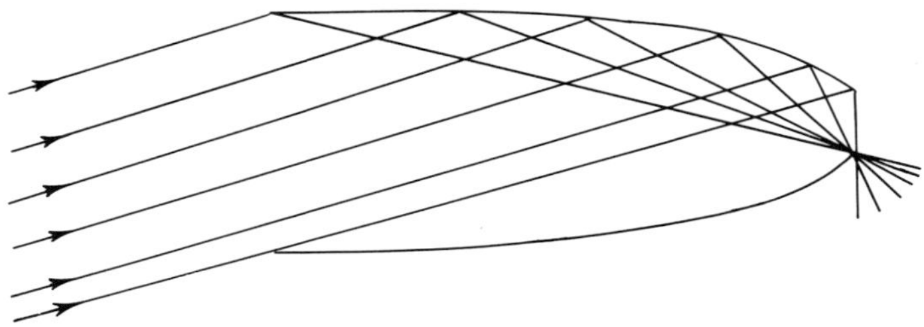

Fig. 9.4 The rays entering a CPC at the extreme angle form a point caustic at one edge of the exit aperture.

the rays will be more spread out, as in Fig. 9.5. Nevertheless, there is marked bunching. As a quantitative example, an ordinary 2D CPC was assumed with a collecting angle of 5° and truncated to give a concentration ratio of 4. Figure 9.6 shows the ray density across the plane exit aperture for pencils at different entry angles—0°, 1.67°, and 3.33°—with different assumed standard deviations σ_θ of the slope error. The very strong effect of a slightly wobbly surface in smoothing the distribution is clearly seen. Figure 9.7 shows the transmission-angle curves for this system and as suggested in Section 9.3, there is little loss in a moderate spreading of the angle of reflection.

As a different example, a 2D system with cylindrical absorber was taken, as in Fig. 9.8. The acceptance angle was 33.75° but the collector was truncated to give a concentration ratio of 1.6 (down from the theoretical maximum value of 1.8). This acceptance angle is appropriate to a totally stationary solar collector aligned in the east–west direction, a case of considerable practical importance. If we trace rays at some angle inside the acceptance, say 15° as in the diagram, we find a very nonuniform distribution of intensity around the absorber.

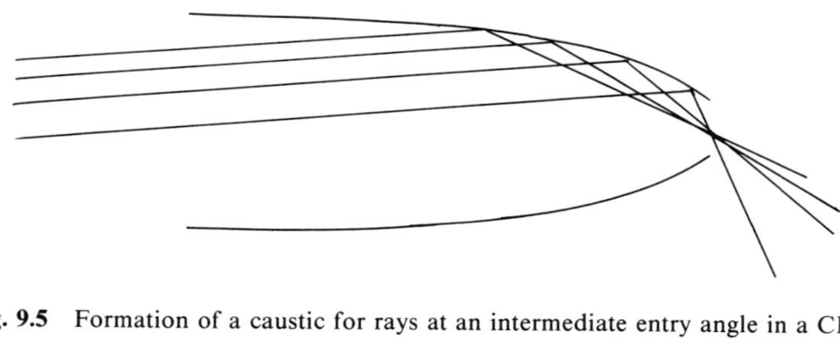

Fig. 9.5 Formation of a caustic for rays at an intermediate entry angle in a CPC.

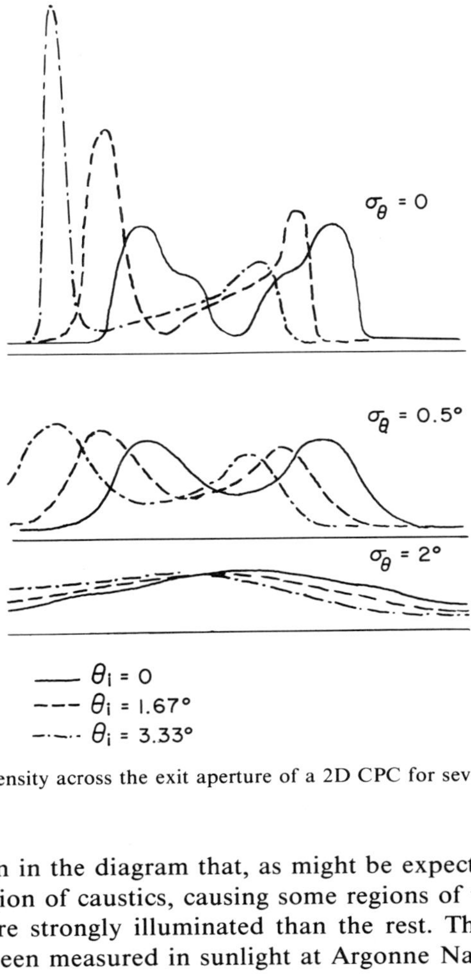

Fig. 9.6 Ray density across the exit aperture of a 2D CPC for several values of σ_θ.

It can be seen in the diagram that, as might be expected, this is due to the formation of caustics, causing some regions of the cylinder to be much more strongly illuminated than the rest. The illumination pattern has been measured in sunlight at Argonne National Laboratory. Figure 9.9 shows the measured peaks falling somewhat lower than calculated (due to slight manufacturing error in the mirror profile and the finite angular subtense of the sun) but nevertheless very pronounced. For certain purposes (see Chapter 10) this can be a great nuisance. It is therefore of considerable interest to see whether deliberately introduced irregularities in the concentrator profile will help. A simple way to introduce irregularities is to subdivide the profile into straight line segments. McIntire at Argonne National Laboratory has

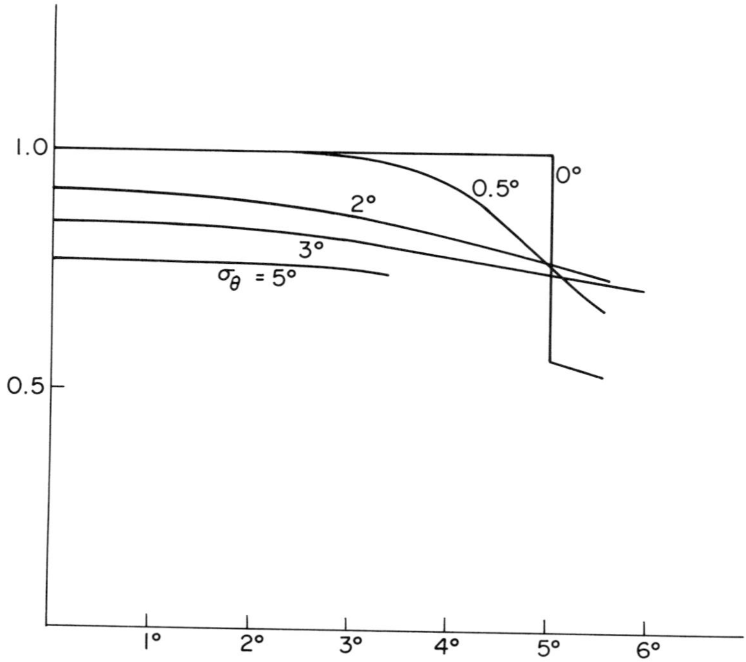

Fig. 9.7 The transmission-angle curves for the systems of Fig. 9.6.

examined the effects of breaking the profile into successively smaller numbers of segments in the context of a stationary solar concentrator (see Chapter 10). Figures 9.10 and 9.11 show that breaking the curve into 20 straight line segments while barely discernible from the exact profile markedly lowers the peaks. A coarser subdivision smooths the illumination even more; of course, as one departs substantially from the ideal profile, the transmission is reduced. Figure 9.12 shows the transmission-angle graph for several cases, the number of segments being indicated on each curve. Other computations not illustrated show that irregularities causing a moderate spreading of the angle of reflection dramatically reduce the sharp maxima with little loss in transmission.

The effect of randomizing the angles of the reflected rays on 3D concentrators is not quite so good as for 2D systems. This is because, even for quite large variances of the angle, pencils incident at angles

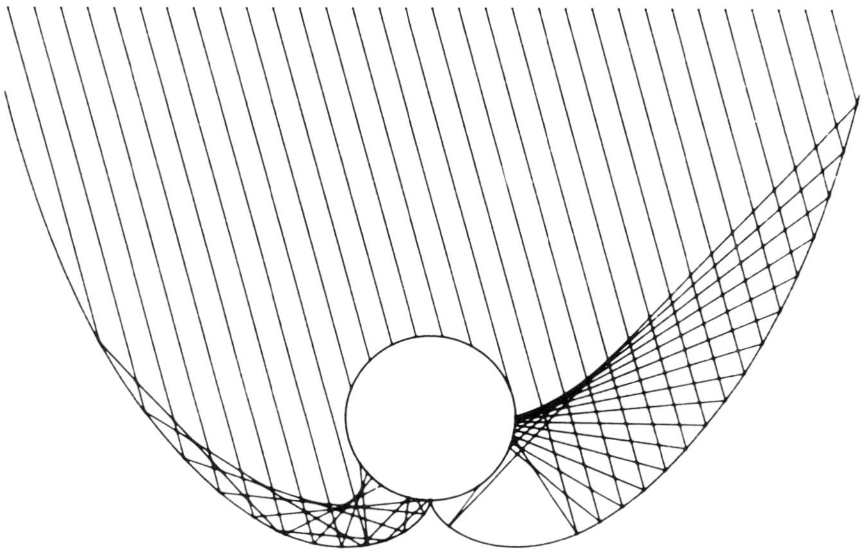

Fig. 9.8 A 2D collector for a cylindrical absorber with acceptance angle 33.75° and truncated to a concentration ratio 1.6. The rays at 15° form sharp peaks in light concentration; the nonuniformity at other angles of incidence is less pronounced.

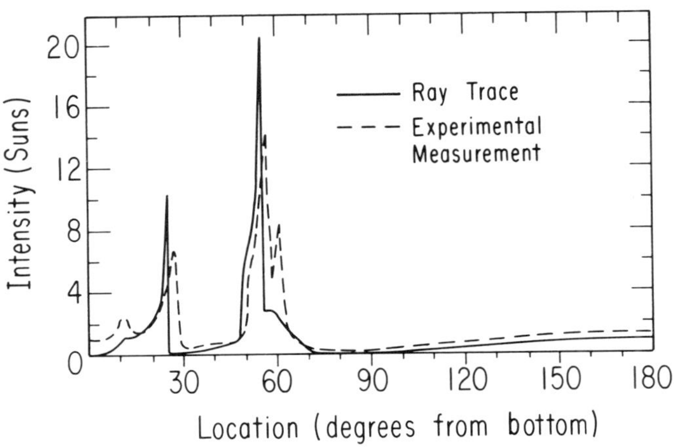

Fig. 9.9 Measured intensity of sunlight around the cylinder; the angle of incidence is normal to the aperture. The angles around the cylinder are measured from the point opposite the reflector cusp.

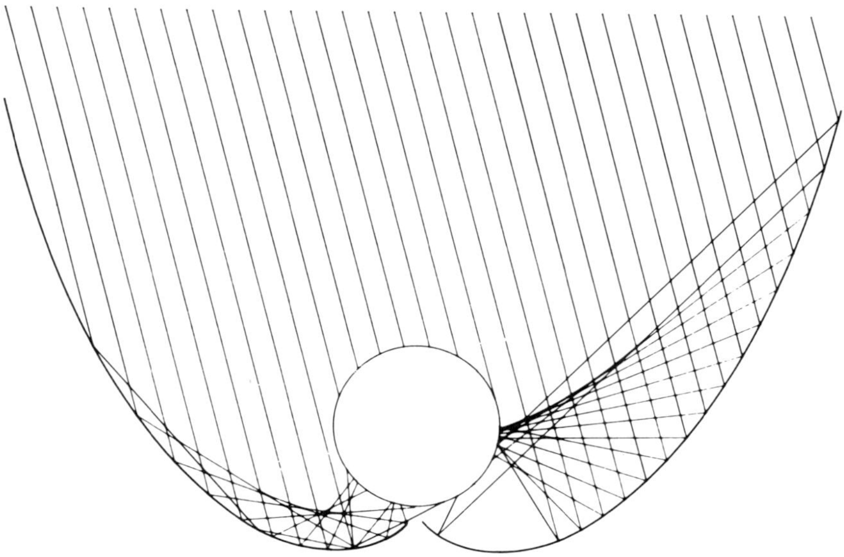

Fig. 9.10 The effect of breaking the concentrator profile of Fig. 9.8 into 20 straight line segments; the peaks from the rays at 15° are suppressed.

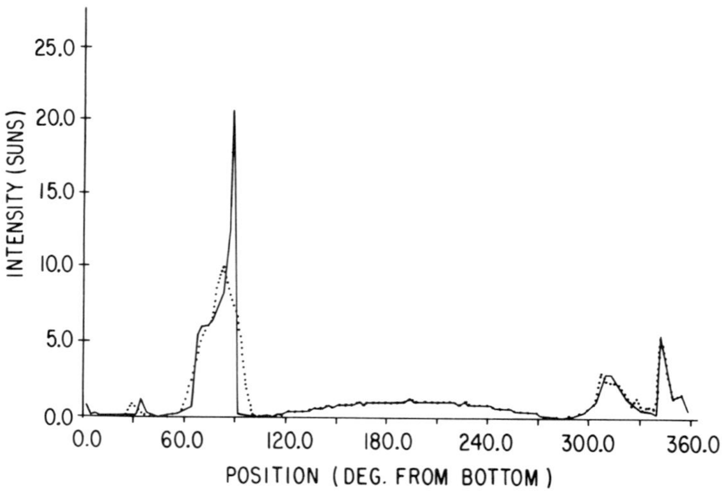

Fig. 9.11 The illumination distribution for the system of Fig. 9.10 showing marked suppression of the peaks; the distribution resulting from the exact profile is superimposed for comparison.

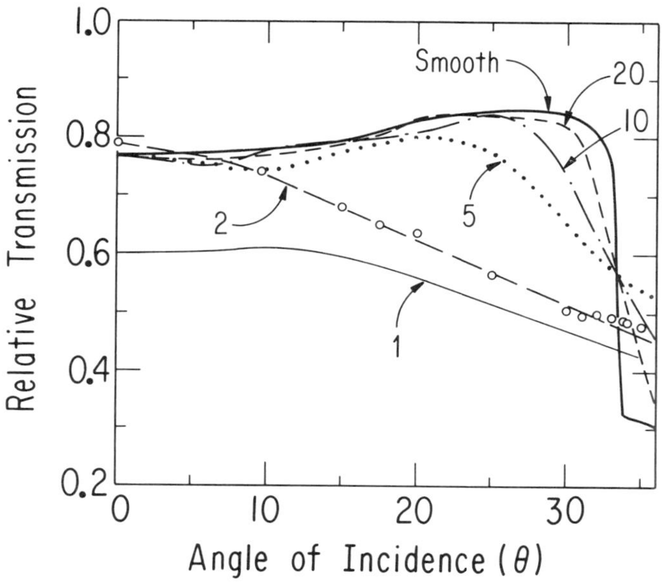

Fig. 9.12 Transmission-angle curves for the concentrator of Fig. 9.8, showing the effect of breaking the concentrator profile into successively smaller numbers of straight line segments. The calculation assumes a mirror reflectivity of 85%, an appropriate angular dependence for the (selective) absorber and a realistic gap between absorber and mirror (see Chapter 10).

much smaller than the extreme angle of the concentrator still give quite a large increased central intensity in the exit aperture. This effect is perhaps understandable in terms of the axial symmetry of the system, since rays randomly deviated toward the axis from all azimuthal points emerge in a region of smaller area than those deviated away from the axis. We show some examples of ray-trace calculations on a 5° 3D CPC truncated to 5% of its original length, giving a concentration ratio of 16 (132 if not truncated). Figure 9.13 shows the relative density of rays emerging from the exit aperture for an axially incident pencil with $\sigma_\theta = 0°$, 0.5°, and 2°, respectively. The percentage of rays transmitted is indicated beside each diagram.

The off-axis pencils benefit more from the angle randomizing, as can be seen from Figs. 9.14 and 9.15, but it would be necessary to go

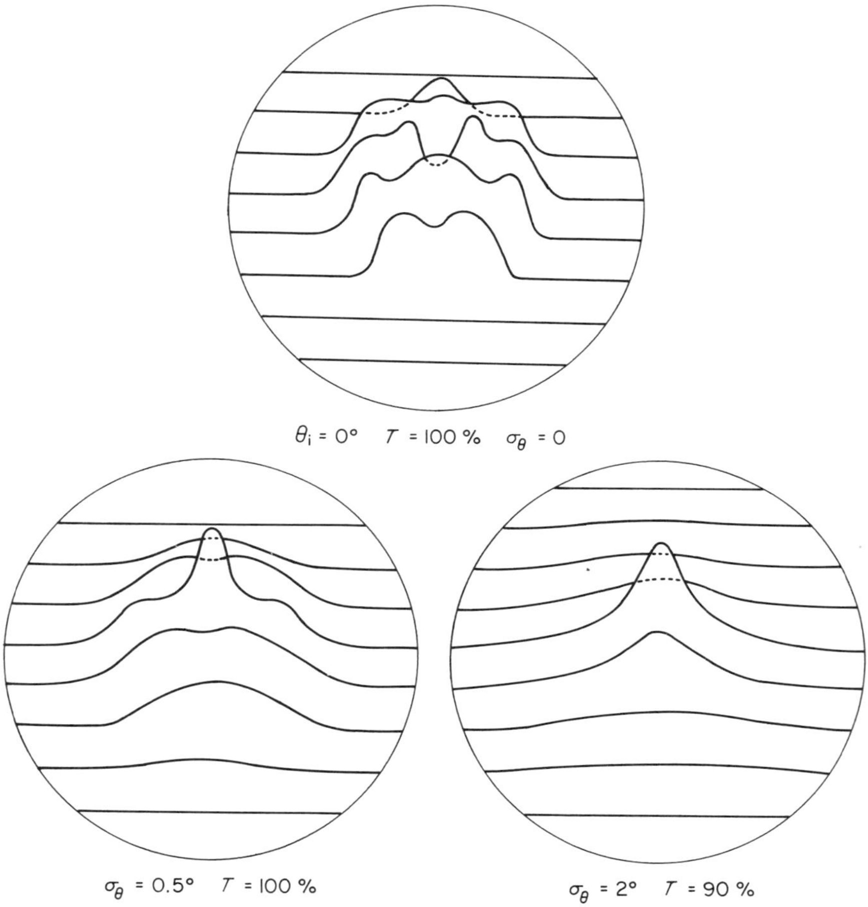

Fig. 9.13 Emerging ray densities for an axial pencil in a truncated 3D CPC (θ, $=5°$, truncated to 5% of original length) with different random slope variances σ_θ^2. The angle deviations in axial planes were governed by a Gaussian probability density with variance σ_θ^2. The angles of the aximuthal or circumferential planes containing these deviated rays were distributed rectangularly over 0 to 2π.

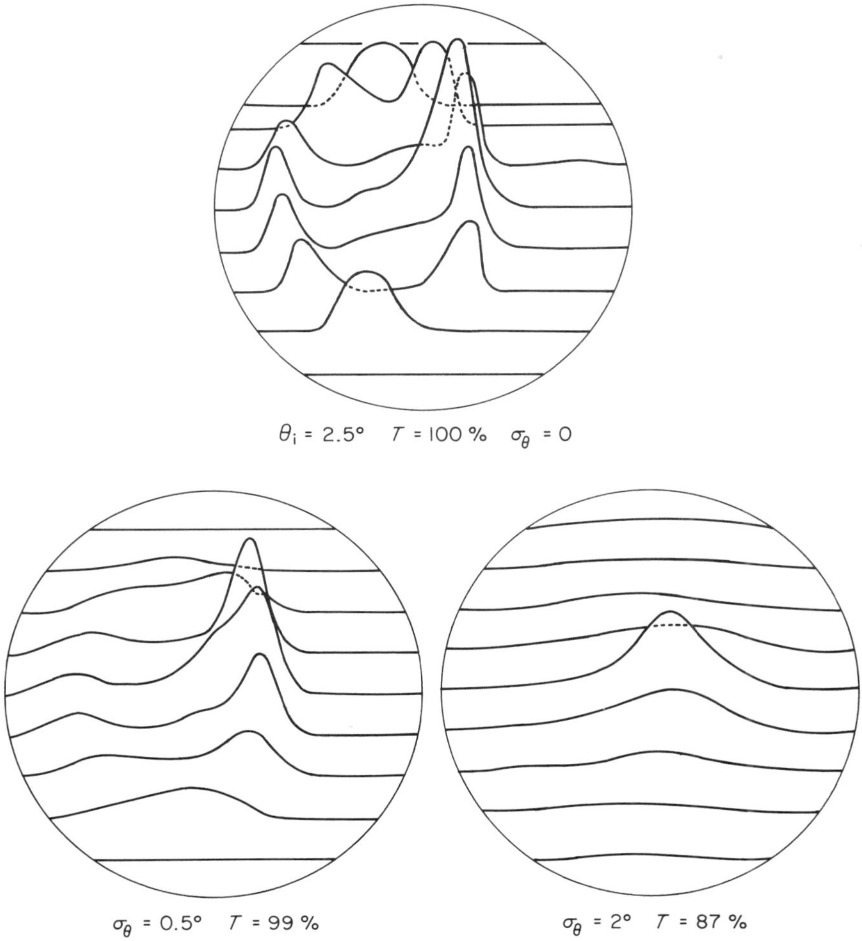

Fig. 9.14 As for Fig. 9.13, but for a pencil at 2.5° to the axis.

to a very large value of σ_θ, about 5°, to get markedly smoothed distributions. Figure 9.16 shows the effect with $\sigma_\theta = 5°$ for an axial pencil and two off-axis angles. However, as might be expected, the transmission losses are quite high in this case and it is probably not worth considering in practice.

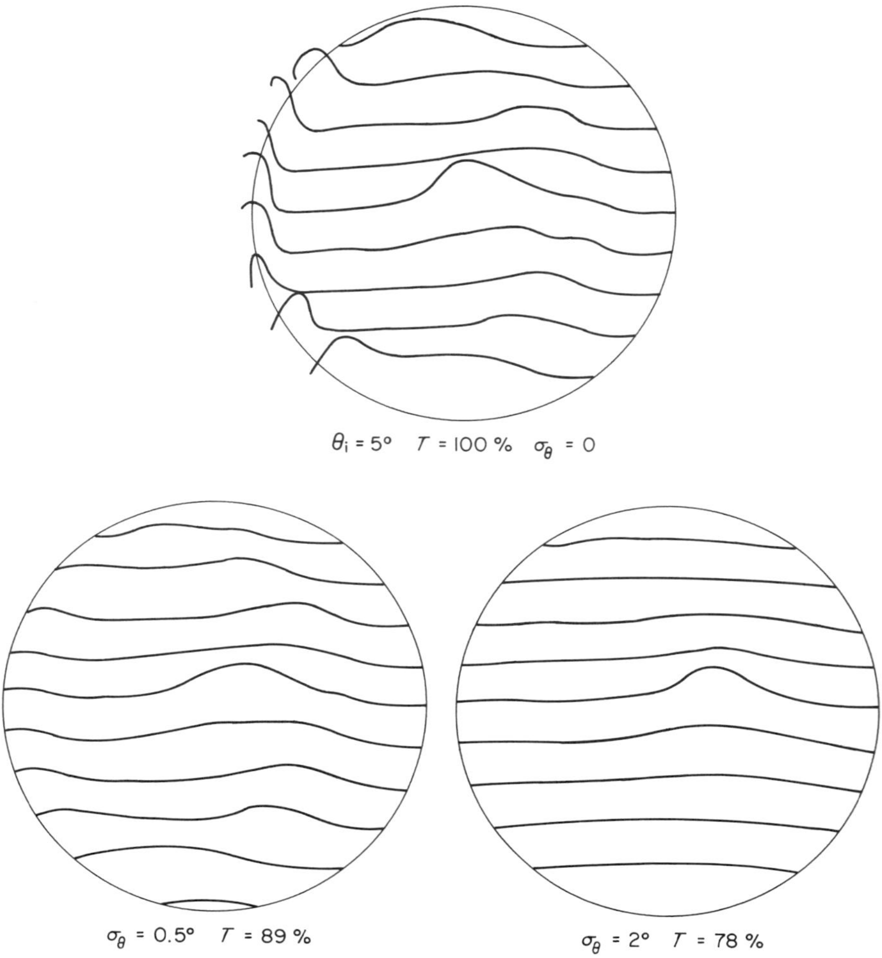

Fig. 9.15 As for Fig. 9.13, but for a pencil at 5° to the axis.

9.5 Reflectors for Uniform Illumination

If we take a single pencil from a point source (or a parallel pencil of light), it is possible to design a reflecting surface that will redistribute the rays uniformly over a chosen output surface (Elmer, 1980). Prob-

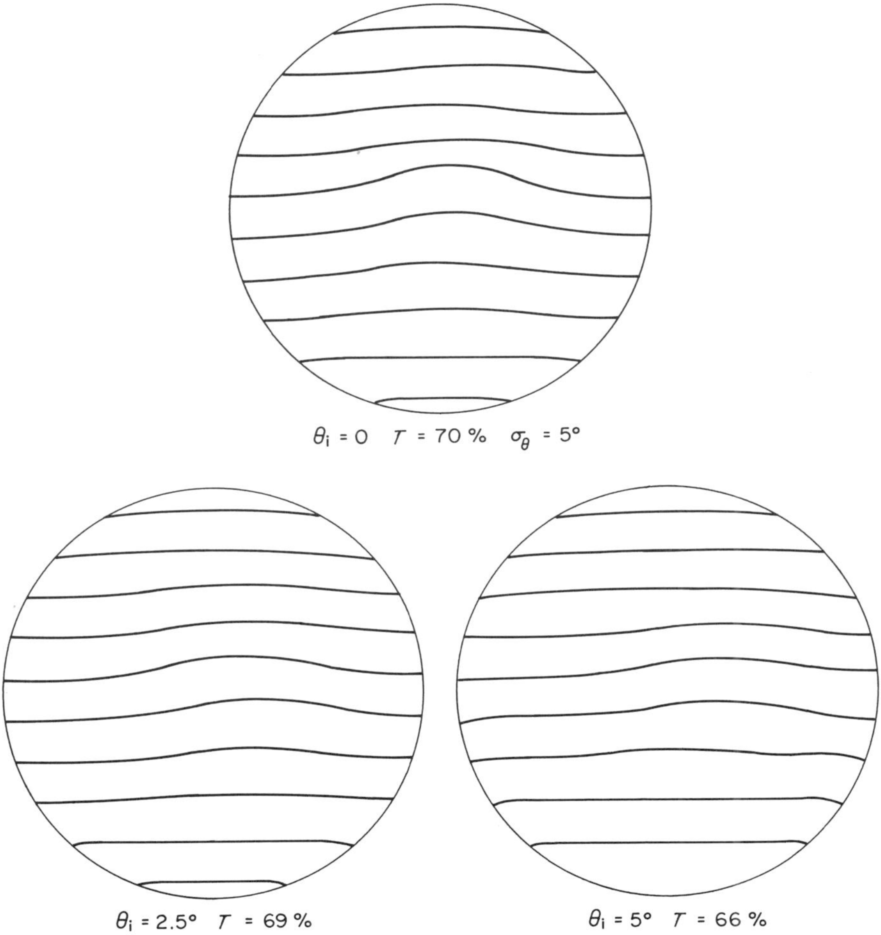

Fig. 9.16 5° concentrator truncated to 5% of its length and with $\sigma_\theta = 5°$, showing the ray densities at the exit aperture for the axial pencil and for 2.5° and 5° off axis.

lems of this kind occur in the design of lighting fixtures, and Burkhard and Shealy (1975), while slanting the discussion toward applications to solar energy, showed how to set up differential equations specifying the shape of the reflecting surface for a variety of output surface shapes.

It is possible to use the methods of Burkhard and Shealy (1975) to design concentrators resembling CPCs in general shape that illuminate

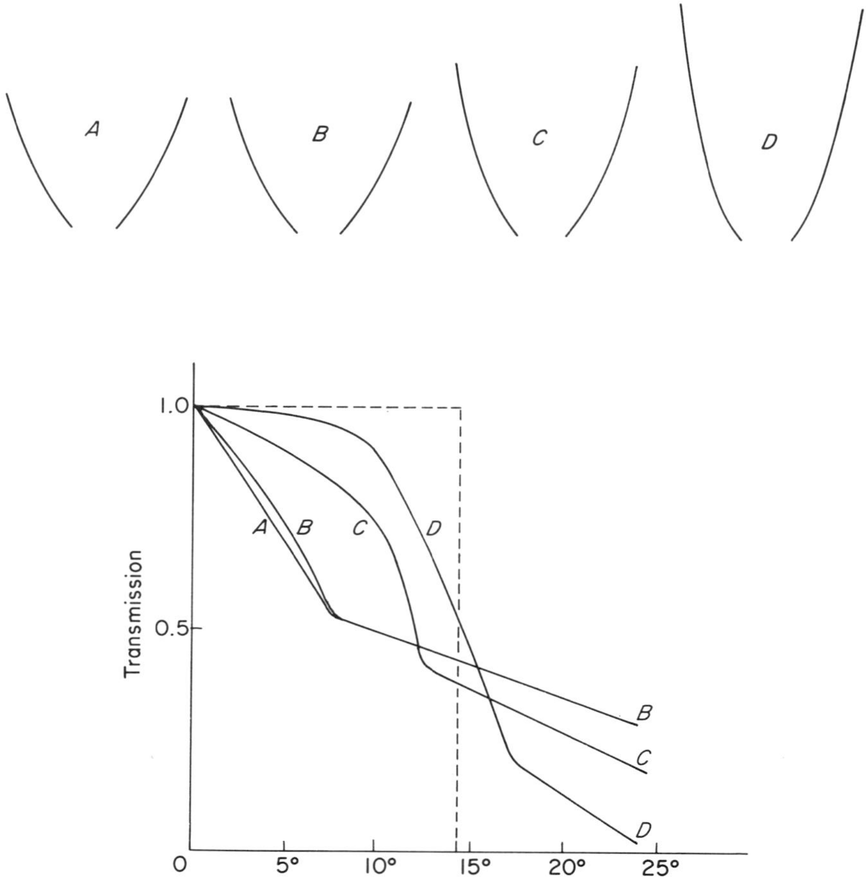

Fig. 9.17 Concentrators based on the Burkhard–Shealy theory. A curve for a 14.5° CPC is superimposed on the transmission-angle curves.

the exit aperture uniformly from an axial pencil. However, if off-axis pencils within the acceptance angle are traced, it is found that much more flux is rejected than with a CPC. For example, four such 2D concentrators were designed with concentration ratio 4* to distribute

* The design is not specified uniquely by the chosen concentration ratio. It depends on a choice of certain rays defining a range of integration, and this yields concentrators of different lengths for the same concentration ratio.

the axial pencil uniformly over the exit aperture, and the transmission-angle curves were calculated. The results are shown in Fig. 9.17. The broken line is what would be obtained from a 2D CPC with concentration ratio 4. The diagram shows also the shapes drawn to scale. Clearly, the Burkhard–Shealy shapes are less efficient as collectors over a finite angle than CPCs.

9.6 Materials and Manufacture

We mention here briefly some of the manufacturing techniques that have been used for nonimaging components. Since the tolerances on shape are relatively mild, ordinary engineering methods rather than the specialized methods of lens and mirror manufacture are generally acceptable. Thus for small-scale purposes CPCs and CPC-like shapes can be turned on a male master and then the hollow component is made by electroforming. Solid components can be turned from plastics materials or, for quantity production, injection molded; since quite high-quality imaging lenses are now made by injection molding plastics, the quality is certain to be adequate for nonimaging optics. Also, molded glass lenses can now be made of reasonable quality and thus for high-temperature applications and quantity-manufacture, molded-glass CPCs, etc., are possible.

Large-scale concentrators for solar work have been made in a variety of ways, of which we mention a few: Spun copper, silver coated, for flow-line concentrators (FLCs); aluminum sheet bent on wooden formers for 2D CPCs; and petallike segments of pressed metal for large primary mirrors to be followed by second-stage CPCs or FLCs. To a large extent, the methods are dictated by local availability of materials and machinery, and it would be pointless to generalize about methods for very large components.

CHAPTER 10

Applications to Solar Energy Concentration

10.1 The Requirements for Concentrators

It may be appropriate to remind ourselves why it is advantageous to concentrate the solar flux. After all, it is well known that flat-plate collectors are adequate for heating a working fluid at temperatures up to about 80°C, and this is at present the most economical way to use solar power for domestic hot water purposes. It is probably also best for domestic space heating in moderate climates, with the reservation that larger glass windows applying the greenhouse effect are also good when appropriate to the use of the building. However, even modest levels of concentration (say, 1.5–2) can dramatically increase the temperature at which heat from a solar collector can be extracted efficiently. The economics of photovoltaic conversion can improve significantly with intermediate levels of concentration (say, 10–100). At high flux levels (of, say, $\approx$1000) efficient generation of electricity is technically feasible; and at the very highest flux levels achievable (say, 40,000 or more) certain processes (e.g., efficient laser pumping) become accessible to solar energy utilization, which may be of considerable interest. Therefore while one may argue over the appropriate flux level for a particular application, there is general agreement that some degree of concentration would be desirable for most applications.

Why then are most practical solar installations nonconcentrating flat plates? The answer must be that the penalties in cost and complexity associated with optical concentration outweigh its advantages. It is the task of nonimaging optics to mitigate these penalties, i.e., to concentrate the solar flux at the least possible "cost" in tracking tolerance, precision of components, and so on.

In effect the optics of solar flux concentration has usually been "nonimaging" since the image quality is rarely an issue. But what nonimaging optics has come to signify is the systematic study and development of design techniques that optimize angular acceptance and throughput efficiency for a given flux concentration, i.e. "high collection optics."

Concentrators are needed wherever we want to use solar power for applications requiring heat at temperatures above about 80°C.* There are numberless applications requiring temperatures up to about 300°C for both domestic and industrial applications, for example, space cooling, cooking, desalination, electricity generation (via steam or some working fluid for turbines). Among the factors that influence the design of concentrators for solar energy are the following:

(1) Cost and ease of manufacture on an appropriate scale
(2) Extent of guidance required for following the sun
(3) Durability and maintenance
(4) Required working temperature of absorber
(5) Preferred geometry of absorber in relation to the mode of utilization of the energy
(6) Susceptibility to contamination and durability under ultraviolet irradiation

Most of these are, of course, interlinked, and their relative weightings are changing in response to economic factors rather rapidly at the time of writing. We cannot, therefore, lay down specific guidance and we restrict ourselves to descriptions of different systems in which we call attention to specific points of advantage or disadvantage.†

* For *evacuated* flat-plate collectors, the upper limit of useful operating temperature is somewhat higher. However, their performance is improved significantly by concentration.

† For a review of solar energy technology see Duffie and Beckman (1980), Kreider and Kreith (1981), and Rabl (1985). Nonimaging applications are discussed by O'Gallagher and Winston (1983).

10.2 Earth–Sun Geometry

It is useful to recapitulate some of the relationships that express the apparent position of the sun relative to a fixed location on the earth. Discussions of this classical subject in terms relevant to solar energy utilization are given by Winston (1974) and Rabl (1976b, 1985). Although the apparent movement of the sun is very familiar, the geometry is, in fact, rather complicated. We are interested in two extreme cases: (1) when a concentrator has to be guided so as to have its axis pointing always directly at the sun; and (2) when the collector is fixed and a compromise has to be reached by using a fairly large entry solid angle. Intermediate cases between these extremes have been proposed for many installations.

To derive the apparent path of the sun, we take coordinates as in Fig. 10.1, where the axes are fixed on the earth. In this diagram the

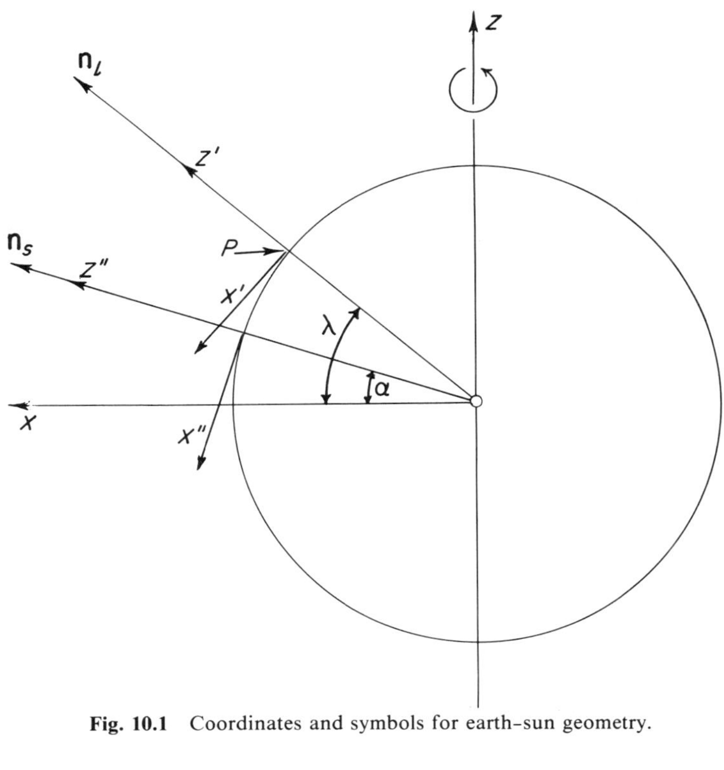

Fig. 10.1 Coordinates and symbols for earth–sun geometry.

z axis points north and the x axis is in the local meridian plane. $\mathbf{n}_s$ is a unit vector pointing to the sun and $\mathbf{n}_l$ is the local vertical at P, the point of observation. We ignore fine points such as the nonspherical shape of the geoid. Let $\mathbf{n}_s$ make an angle α with the equatorial plane at noon and let the latitude of P be λ. It is easily shown that

$$\sin \alpha = -\sin E \cos(\Omega D) \tag{10.1}$$

where $E = 23°27'$, $\Omega = (2\pi/365.25)\text{ days}^{-1}$, D = time in days after winter solstice, again ignoring ellipticity of the orbit, etc.

Next we can show that $\mathbf{n}_s$ is given by

$$\mathbf{n}_s = (\cos \alpha \cos \omega t, -\cos \alpha \sin \omega t, \sin \alpha) \tag{10.2}$$

where

$$\omega = (2\pi/24)\text{ hours}^{-1}, \qquad t = \text{time in hours after noon}$$

Now let

$$(k_{x'}, k_{y'}, k_{z'}) = (\sin \lambda \cos \alpha \cos \omega t - \cos \lambda \sin \alpha,$$
$$-\cos \alpha \sin \omega t, \cos \lambda \cos \alpha \cos \omega t + \sin \lambda \sin \alpha) \tag{10.3}$$

which gives the direction of the sun as seen from P with the time t as parameter. The path is more easily comprehended if we express it in $k_{y'}$, $k_{z'}$ space by eliminating t:

$$\left(\frac{k_{z'} - \sin \lambda \sin \alpha}{\cos \lambda \cos \alpha}\right)^2 + \left(\frac{k_{y'}}{\cos \alpha}\right)^2 = 1 \tag{10.4}$$

This is an ellipse with its center at $[(1 - \sin^2 \lambda \sin^2 \alpha)^{\frac{1}{2}}, 0, \sin \lambda \sin \alpha]$ and with major and minor axes $\cos \alpha$ and $\cos \lambda \cos \alpha$, i.e., it has eccentricity $\sin \lambda$. Figure 10.2 shows such ellipses plotted for a latitude of 40° and for three different times of year. This representation is perhaps reasonably near to one's subjective impression of the movement of the sun, since the $k_{y'}$ axis corresponds to the southern (in the northern hemisphere) horizon. But there are other representations that are more useful for the design of concentrators. If we change the axes so that z'' is along the noon direction of the sun at any given time of year, y'' is east–west as before, and x'' is perpendicular to these two (Fig. 10.1), we easily find the following equation between the x'' and y'' direction cosines:

$$\left(\frac{k_{y''}}{\cos \alpha}\right)^2 + \left(\frac{k_{x''} + \sin \alpha \cos \alpha}{\sin \alpha \cos \alpha}\right)^2 = 1 \tag{10.5}$$

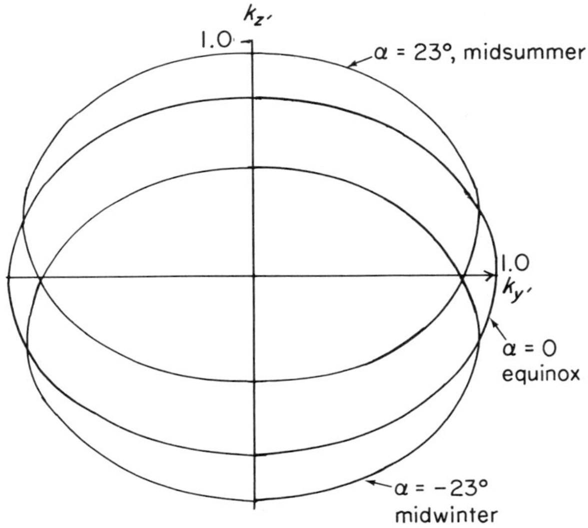

Fig. 10.2 The path of the sun in $k_{y'}$, $k_{z'}$ space. These direction cosines correspond to east-west and local zenith axes. The origin is in the direction of the south point on the horizon. The line $k_{z'} = 0$ represents the horizon.

This is again an ellipse, with axes $\cos\alpha$ and $\frac{1}{2}\sin 2\alpha$ and with center at $[1 - \sin\alpha\cos\alpha,\ 0,\ (1 - \sin^2\alpha\cos^2\alpha)^{\frac{1}{2}}]$ so that it has eccentricity $\cos\alpha$. This representation, which is independent of the latitude, is shown in Fig. 10.3. Representing the solar motion in direction cosines is useful in visualizing the match to stationary collectors since, as discussed in Chapter 5, the angular acceptance of an ideal 2D concentrator is also an ellipse when expressed in these variables.

There is yet another representation that is particularly useful in discussing 2D concentrators such as CPCs that are held in fixed orientation (i.e., due south) with the possibility of some adjustment in elevation. We imagine a plane through the y axis of Fig. 10.1 and we ask what angle this makes with the equatorial plane of the earth if it is arranged always to contain the sun as it passes across the sky. This angle φ_v, which may be called the solar elevation (Rabl, 1976b), is a measure of the elevation adjustment that would be required in a 2D CPC if it had to be adjusted so that its plane of symmetry always contained the sun. Alternatively, the change in φ_v is a measure of the

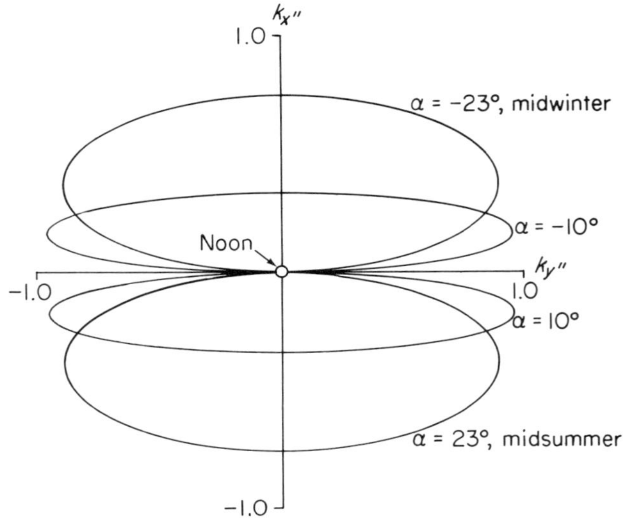

Fig. 10.3 The path of the sun in $k_{x''}$, $k_{y''}$ space: the y'' axis is east–west and the x'' axis is perpendicular to the plane of the ecliptic. Thus the origin gives the position of the sun at noon. In this diagram the ellipse has the same shape for all latitudes at a given time of year.

acceptance angle required of a CPC with no elevation adjustment. It is not difficult to show that φ_v is given by

$$\tan \varphi_v = \tan \alpha / \cos \omega t \tag{10.6}$$

In Fig. 10.4 this function is plotted against the number of hours before or after noon for different numbers of days before and after the summer solstice. On this figure the direction of the midplane of a 2D CPC is represented by a horizontal straight line at the appropriate level. Similarly, two horizontal straight lines spaced apart $2\theta_i$ represent the acceptance of a 2D CPC of entry angle θ_i. We can therefore use this graph to estimate the proportion of useful hours that a CPC will have if set at any chosen angle. For example, the broken lines correspond to $\theta_i = 30°$ with the concentrator direction set toward noon sun at equinox. To be more precise, we should allow also for the finite angular subtense, 0.5°, of the sun and perhaps also for slope errors in the concentrator. We can also use Fig. 10.4 to estimate the gains in using higher concentration ratios and making seasonal adjustments to the concentrator elevation.

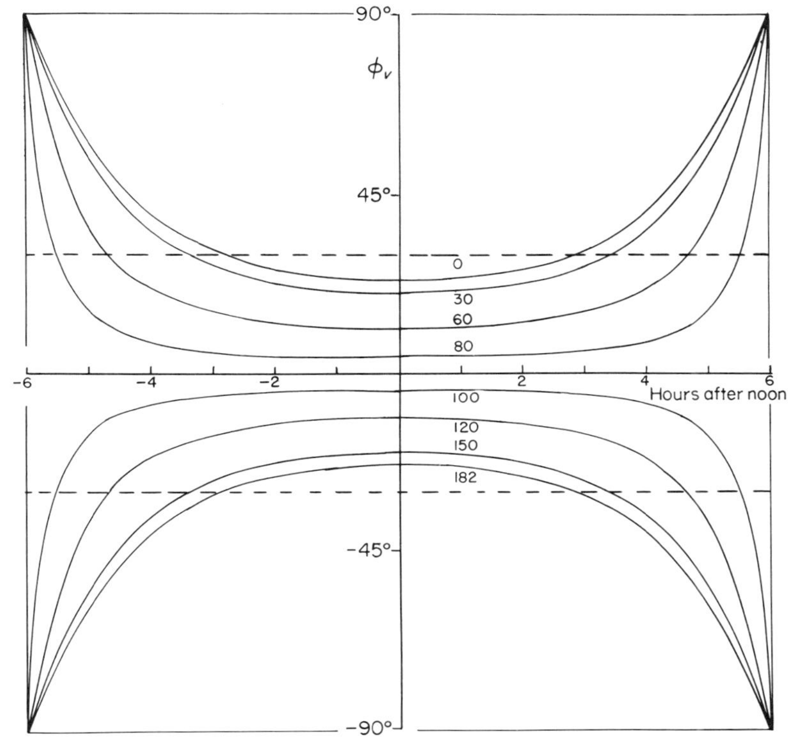

Fig. 10.4 The elevation angle ϕ_v of the sun. This is the angle of the sun above or below the plane of the equator at different times of day [Eq. (8.6)]. Horizontal lines can be drawn on this graph representing the extreme entry angle of a 2D CPC, so as to show the times during which direct sunlight is collected. Each curve is labeled with the number of days after midsummer.

10.3 Insolation Characteristics

It is important to recognize that we have been considering the incident solar energy flux (or, as it is sometimes called, the *insolation*) as entirely coming from the direction of the solar disk, i.e., the direct beam radiation. The atmosphere attenuates the insolation by both absorption and scattering. Aside from causing the familiar changes in intensity associated with variations in cloudiness and weather, this results in a significant component of insolation outside the solid angle of the solar disk, called the *diffuse radiation*. Were this diffuse radiation isotropic, then the fraction of it collected by a concentrator would be at most the reciprocal of the concentration ratio independent of design details. However, the diffuse component generally peaks toward the solar disk. The degree of anisotropy is particularly pronounced in the proximity of industrial and built-up areas, where dust and aerosols are present to increase the turbidity of the atmosphere. To illustrate, data for Mainz, Federal Republic of Germany (Fig. 10.5), show the anisotropy extending to relatively large angles from the solar disk. Such "diffuse" radiation can be concentrated. This favors the use of concentrators with large acceptance angle.*

10.4 Collector Design

The compound parabolic concentrator (CPC) is not a specific collector, but a family of collectors based on a general design principle for maximizing the geometric concentration, C, for radiation within a given acceptance half-angle $\pm\theta_c$. This maximum limit exceeds by a factor of 2 to 4 that attainable by systems using focusing optics. The wide acceptance angles permitted using these techniques have several unique advantages for solar concentrators, including the elimination of the diurnal tracking requirement at intermediate concentrations (up to $\approx 10\times$), collection of circumsolar and some diffuse radiation, and relaxed tolerances. Because of these advantages, CPC-type concentrators have applications in solar energy wherever concentration is desired, e.g., for a wide variety of both thermal and photovoltaic uses. Selected configurations for both nonevacuated and evacuated thermal

* The nondirect radiation that appears, to an observer on the ground, to originate from the region around the sun is sometimes designated "circumsolar," while the term "diffuse" is reserved for the larger angle component. This radiation is caused by scattering of light through small angles by particles (aerosols) in the earth's atmosphere with dimensions on the order of or greater than the wavelength of light. The distinction, while not fundamental, may be useful in practice because the instrument normally used to measure the direct beam of radiation, the pyrheliometer, has a 5–6° (full angle) field of view as compared to the $\frac{1}{2}°$ subtended by the sun. The pyrheliometer thus includes within its acceptance cone a substantial fraction of the circumsolar radiation as well as the direct beam radiation. Nevertheless, the usual procedure for estimating "diffuse" radiation is to subtract the pyrheliometer measurement from the total radiation falling on a plane normal to the sun.

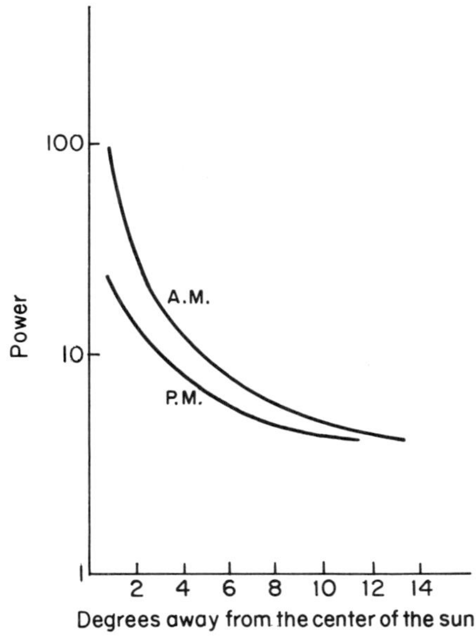

Fig. 10.5 Insolation at Mainz, Federal Republic of Germany. The graph shows the power density received at ground level in a spectral band 13 nm wide centered at 448 nm wavelength, from the sky at the stated angular distances from the center of the sun's disk on a typical autumn day. For comparison, the ordinate inside the disk of the sun would be of the order of 10^5 on the same scale. Thus the diffuse radiation if collected would amount to 20–30% of the direct solar radiation, as has been verified experimentally. Adapted from Eiden (1968).

collector applications are discussed with particular emphasis on the most recent advances. The use of CPC-type elements as secondary concentrators is illustrated in the context of higher concentration photovoltaic applications and ultra-high flux solar furnaces for solar laser pumping.

The introduction of the concept of nonimaging optics for solar concentrators can be traced to a system for coupling photomultipliers to Čerenkov counters and other extended sources (Hinterberger and Winston, 1966a) illustrated in profile in Fig. 10.6. Essentially in designing a concentrator as an image-forming system, unnecessarily high standards are being set since an image of the sun is not needed; all that is needed is that the rays collected shall all fall within as small

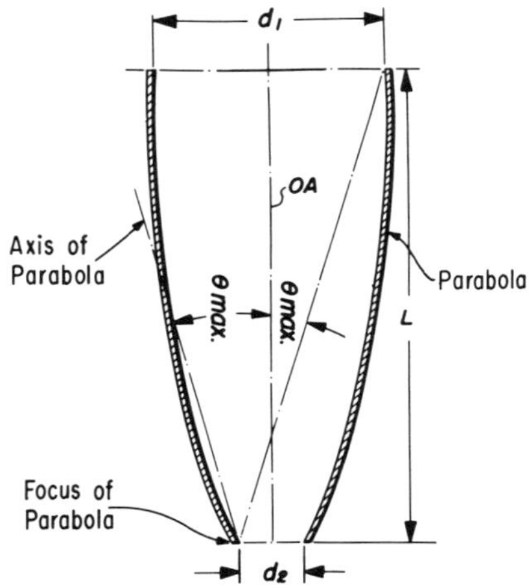

Fig. 10.6 Cross-sectional profile of nonimaging cone concentrator. The device collects radiation entering the circular entrance aperture of diameter d within angle of $\pm\theta_{max}$ and directs it to the smaller diameter aperture d_2 with negligible losses, and thus closely approaches the thermodynamics limit.

an area as possible, on the proposed absorber surface. In order to achieve this, the elaborate edifice of aberration theory used by conventional optical designers can be replaced by a single key idea, the edge-ray principle. This is the statement that the most efficient concentrator for a given concentration ratio is that which makes all rays from the extreme input angle (e.g., $\frac{1}{4}^{\circ}$ in the case of the sun) fall on the edge of the absorber. Concentrators which can be designed to satisfy this principle attain or very closely approach the maximum permissible concentration for a given angular field of view: the so-called thermodynamic limit. Stated most generally, this limit is related to the half-angle of acceptance $\pm\theta_C$ by

$$C_{max} = (\sin\theta_C)^{-1} \text{ (2D geometry)} \tag{10.7}$$

or

$$C_{max} = (\sin\theta_C)^{-2} \text{ (3D geometry)} \tag{10.8}$$

More recently these limits have been shown to apply to systems where wavelength-dependent effects are important (Welford and Winston, 1982) e.g., holographic optical elements.

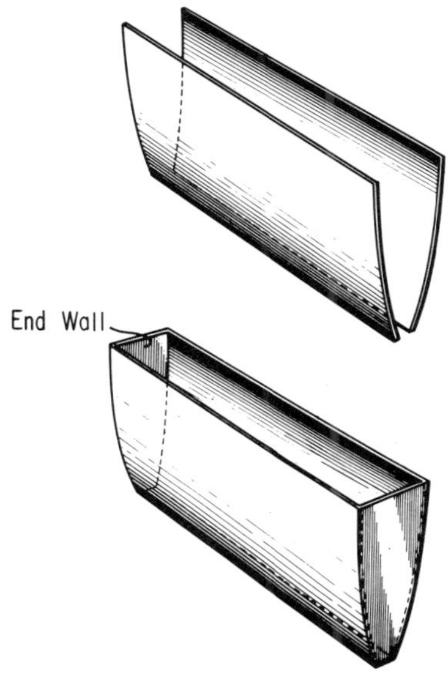

Fig. 10.7 2D or trough concentrator, having the same profile as Fig. 10.6 aligned with its y axis in the east–west direction, has an acceptance field of view well matched to solar motion and can achieve moderate concentration with no diurnal tracking. Vertical reflecting and walls effectively recover shading and end losses.

The application to solar energy of the Čerenkov coupler was suggested by Winston in 1973 and a considerable amount of solar applications development was carried out at Argonne National Laboratory. Significantly, it was recognized that the 2D form of the Čerenkov coupler had certain nonobvious advantages for solar work. The 3D form, i.e., with the profile of Fig. 10.6 rotated about its axis of symmetry, is capable of high concentration ratios but must be guided to face the sun with a precision better than the angular acceptance. A 2D version, looking like a trough with parabolic cylinder sides appropriate to a flat absorber, as illustrated in Fig. 10.7, with the long axis aligned in the east–west direction, can achieve moderate concentration (up to about $10\times$) with no diurnal tracking and even useful concentration (up to about $2\times$) with year-long fixed position.

Modifications of the 2D design for various nonplane absorbers were made by Winston and Hinterberger (1975). Several of the most common solutions recommended for thermal applications are shown in Fig. 10.8 and illustrate the wide acceptance angles possible. Reflecting troughs with these profiles achieve the limit of Eq. (10.7), collecting all rays within $\pm\theta_C$ at the entrance aperture and directing them to the absorber be it a flat surface, fin, wedge, or tube. The solution for a

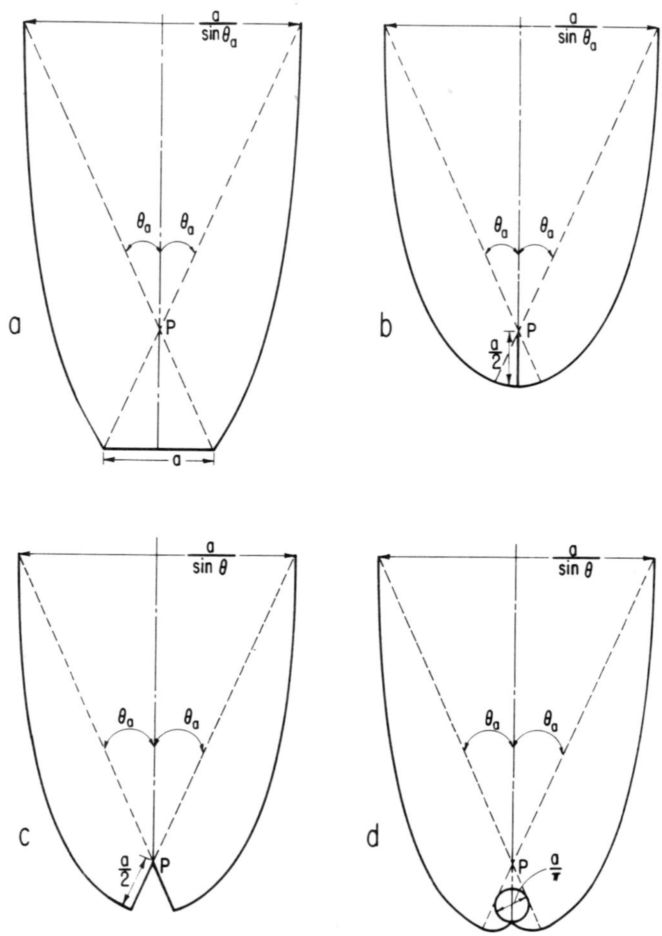

Fig. 10.8 Cross-sectional profiles of ideal trough concentrators generalized to absorbers of different shapes. In practice, the reflectors are usually truncated to about half their full height to save reflector material with only negligible loss of concentration.

flat absorber (Fig. 10.8a) consists of two parabolic segments, each with its axis tilted at an angle θ_C relative to the collecting aperture normal and its focus at the lower edge of the opposite mirror. For the fin and tube solutions (Fig. 10.8b and d) the profile consists of an involute of the absorber inside the "shadow lines": defined by the extreme rays tangent to the absorber at $\pm\theta_C$ and a curve outside the shadow lines whose slope at each point is defined by the condition that a ray parallel to the extreme ray is reflected such that it is just tangent to the absorber. For the fin this corresponds to two tilted parabolas outside the shadow lines and a circular arc segment inside. These all became known as the CPC, although for nonplane absorbers the cylinder profile is frequently nonparabolic. The edge-ray principle proved elastic enough to accommodate most nonplane absorbers of interest and even hybrid systems employing Fresnel lenses (Collares-Pereira *et al.*, 1977).

The classical edge-ray designs require the reflector and absorber to touch. However, this may be undesirable for practical reasons such as the need to thermally isolate the absorber from its surroundings. In response to this, designs have been found which permit gaps between the absorber and the reflector. There are two classes of solutions. In one, the CPC reflector is designed for a virtual absorber as in Fig. 10.9. This is called the "extended cusp design" (Winston, 1978b). Alternatively, the absorber is immersed in a cavity which reflects all radiation back to the absorber as in Fig. 10.10 (Winston, 1980). Simple forms of the cavity can be W shaped (McIntire, 1980) or even V shaped. A good compromise is a V shape with a gap which

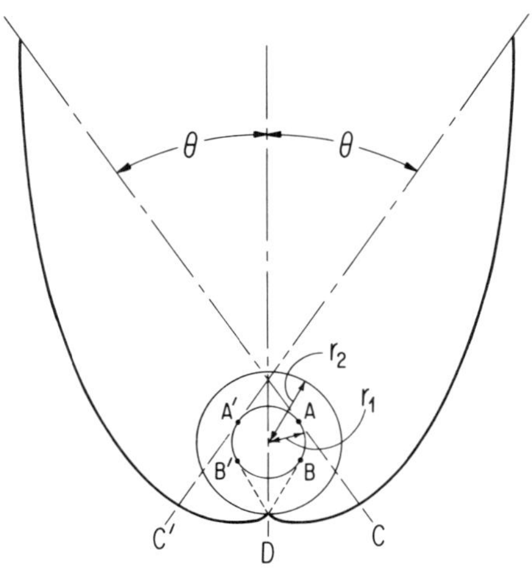

Fig. 10.9 Concentrator designed for virtual absorber ($DB'A'ABD$). This recovers maximum flux concentration on the absorber at the expense of oversizing the reflector.

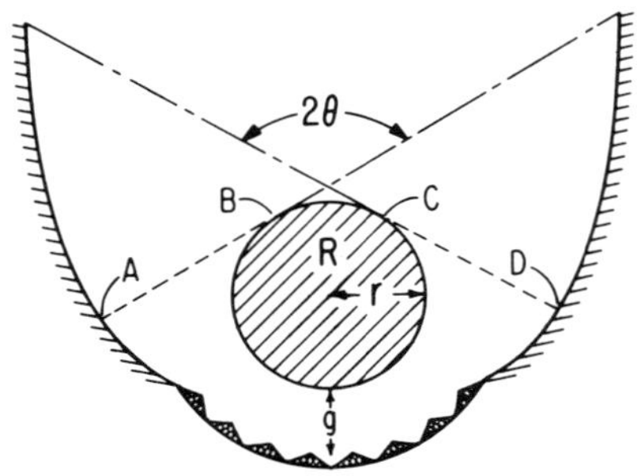

Fig. 10.10 Concentrator for a cylindrical receiver using a grooved cavity. For the example chosen, $g = 0.8r$, $2\psi = 118°$. The angular acceptance $\theta = 60°$ and overall concentration $C = 1$.

permits very small optical loss, typically <2% (McIntire and Winston, 1981).

In recent years the research has centered around the two objectives: (1) the evaluation of design and construction methods for CPCs with nonevacuated absorbers and (2) the development of evacuated CPCs. It was demonstrated that nonevacuated CPCs, although clearly not classified as high-temperature collectors, have a measured performance which exceeds that for typical flat-plate collectors at temperatures above roughly 50°C and should have economic advantages as well if properly constructed. As part of a prototype-development effort, a comprehensive set of design methods and test procedures for nonevacuated CPCs was developed. Nonevacuated CPCs with their relatively high heat losses usually require higher concentrations and some seasonal adjustments (two to ten times annually). Evacuated CPCs can achieve excellent performance with low to moderate concentration, so that fixed-mount totally stationary designs are possible. In characterizing the thermal performance of CPCs in general we describe the thermal collection efficiency η, as a function of operating temperature T by

$$\eta(T) = \eta_0 - U(T - T_a)/I - \sigma\varepsilon(T^4 - T_a^4)/(CI) \tag{10.9}$$

where C = geometric concentration ratio, T_a = ambient temperature, I = total isolation, σ = Stefan–Boltzmann constant, ε = absorber surface emittance, and U = linear heat loss coefficient.

For all nonevacuated CPCs and some evacuated versions the radiative heat loss component is incorporated into the linear term and the third term is dropped. The ambient temperature intercept η_0 sometimes called the "optical efficiency" includes optical losses in transmission through any glazings, reflections, and absorption as well as a correction for loss of some diffuse sky ratiation.

10.5 Nonevacuated CPCs

The design problems for nonevacuated CPC collectors are entirely different from those for CPCs with evacuated absorbers. The former are particularly vulnerable to high heat losses if one uses an improper design. The recommended design constraints are summarized below.

1. Use of one of the "backless" or "wrap-around" designs such as the cusp for a tubular absorber or vertical-fin design of Fig. 10.8b or d
2. Use of selective coatings on the absorber
3. Avoidance of the use of cavity absorbers since they are characterized by high heat losses
4. Minimization or elimination of heat loss via conduction through the reflectors [this can be accomplished by using reflectors whose thickness is negligible compared to the overall dimension (e.g., height, aperture) of the trough, such as metallized plastics or films and by thermally decoupling the absorber from the reflectors by a small gap maintained by insulating stand-offs]
5. Use of a CPC profile calculated for an imaginary absorber surface which is inside the actual absorber surface (a resulting oversize of the actual absorber size of 5–20% is recommended)

Two prototype nonevacuated CPC designs were constructed at the University of Chicago with different objectives in mind. An experimental CPC for a tubular absorber with a very narrow acceptance angle (±6.4°) was built to examine the behavior of a high-concentration small-acceptance-angle CPC. The absorber was a 7.9-mm-diameter copper tube coated with black chrome. The design acceptance angle was ±6.4°, corresponding to an ideal concentration of $C = 9.0\times$ but the profile was truncated to $\approx\frac{1}{3}$ of its full height resulting in a net concentration of $6.5\times$ relative to its design absorber diameter of 6.4 mm and $5.2\times$ relative to the actual oversized absorber used. The full collector consisted of 7 troughs, each 1.8 m long with a total area of 1.67 m^2. Because of its small acceptance angle, it requires 24 yearly

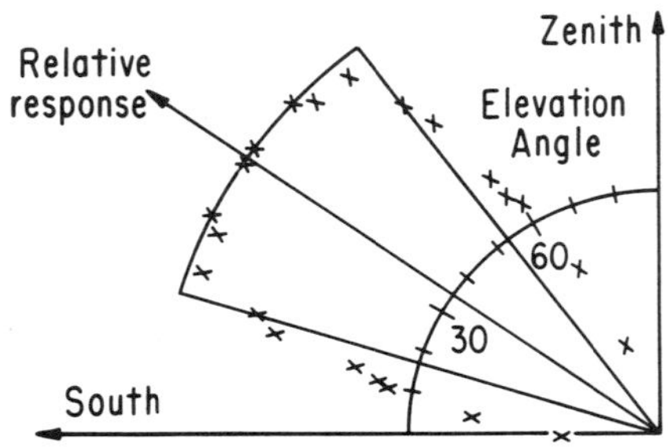

Fig. 10.11 The "ideal" and measured angular response function for the $3\times$ non-evacuated CPC installed at the Navajo school project. The latitude is about 37° at this location so the collector "sees" the sun during the winter only. For year-round collection, two tilt adjustments would be necessary.

adjustments. The performance of this collector was well described by $\eta_0 = 0.68 \pm 0.01$ and $U = 1.85 \pm 0.1$ W/m^{2}K.

A second prototype with $C = 3\times$ for a vertical-fin absorber was built to verify the production design for an experimental array of low-temperature CPC collectors at the Bread Springs School, an elementary school on the Navajo Indian Reservation near Gallup, New Mexico. Since this collector was designed for heating applications only, the acceptance angle was chosen to be ±18° and the collector tilted relative to the zenith by the latitude angle plus 18° as shown in Fig. 10.11. This insures collection for at least 7 hours a day for the 6-month period between the fall and spring equinoxes, with no tilt adjustments. For year-round collection, two tilt adjustments per year would be required. The absorber is a fin (configuration of Fig. 10.8b) with a center tube and each collector consists of a two-trough module 1.4 m^2 in net area and the troughs are 460 mm deep. A Kinglux aluminum sheet 0.5 mm thick is used as the reflector ($\rho = 0.84$). The outer enclosure was constructed of a glass fiber epoxy with water white glass as a cover. Measured performance yielded $\eta_0 = 0.61 \pm 0.03$ with $U = 2.7 \pm 0.2$ W/m^{2}K. At the present time, the nonevacuated CPC is being deployed on a fairly large scale by a company in Israel (Gordon, 1988) for temperatures in the 80–90°C range (Fig. 10.12).

The efficiencies of these nonevacuated CPCs are to be compared with typical values for flat plates with η_0 from 0.70 to 0.78 and U

Fig. 10.12 Field of nonevacuated CPCs in Israel.

from 4.5 to 7 W/m^2K. Despite lower optical efficiencies, the CPCs outperform typical flat-plate collectors above temperatures of as low as 10°C above ambient (for the 5.2×) to about 35°C above ambient (for the 3×). This is particularly important when it is recognized that the 3× should represent a relatively inexpensive collector design. While a detailed economic analysis cannot be based on the prototype construction methods used here and is beyond the scope of the present work, there are several unique features which contribute to its low cost potential, among them the relatively small absorber cost and very limited insulation requirements. In fact, for the application in Israel, the nonevacuated CPC was selected over flat-plate collectors because it was both more efficient and more cost effective.

10.6 Evacuated CPCs

Development work at Argonne National Laboratory after 1975 was strongly influenced by the emergence of the dewar-type evacuated absorber. This thermally efficient device, developed by two major U.S. glass manufacturers, General Electric and Owens Illinois, could, when coupled to CPC reflectors, supply heat at higher temperatures than flat-plate collectors while retaining the latter's advantage of a fixed mount. For example with external anodized aluminum reflectors designed for an acceptance angle of $\pm 35°$ and truncated to a net geometric concentration of between $1.1\times$ and $1.4\times$ such concentrators attain optical efficiencies of between 0.52 and 0.62 depending on whether a cover glass is used. The thermal performance is excellent with values of $\varepsilon \approx 0.05$ and $U \approx 0.5\ \mathrm{W/m^2K}$ in Eq. 10.9. Other versions have opened the acceptance angle to $\pm 50°$ with $C \approx 1.1\times$ to allow polar orientation and are characteried by $U = 1.3\ \mathrm{W/m^2K}$ (lumping the radiative losses in the linear term). An illustration of the basic configuration is shown in Fig. 10.13.

Several manufacturers introduced collectors of these types for applications ranging from heating to absorption cooling to driving Rankine Cycle engines. A number of installations, each in excess of $1000\ \mathrm{m^2}$, have been successfully deployed. Examples include a "SOLERAS" (joint U.S./Saudi Arabia) project in Phoenix, Arizona, driving a

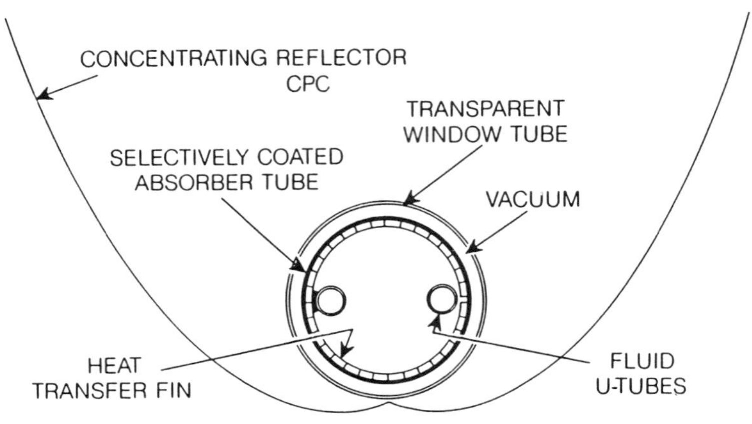

Fig. 10.13 Cross-sectional profile of contemporary evacuation tube CPC according to the basic design developed at Argonne National Laboratory.

Fig. 10.14 Evacuated tube CPC array installed in Phoenix, Arizona, as part of Soleras project.

Rankine engine (Fig. 10.14); the Holiday Inn, Virgin Islands, array for cooling and desalination (Fig. 10.15); and the Illinois Department of Agriculture headquarters' array for heating and cooling in Springfield, Illinois (Fig. 10.16). The concentrators are arranged in panels resembling flat-plate collectors. There are two methods for extracting the heat: Either the fluid is simply allowed to contact the glass, being introduced into the dewar by a feeder tube, or it flows through a closed piping network with metal heat exchange fins actually contacting the inner glass wall (Fig. 10.13).

10.7 An Advanced CPC: The Integrated Concentrator

The optical efficiency of evacuated CPC solar collectors can be significantly improved over that of contemporary commercial versions

Fig. 10.15 Polar-oriented evacuated tube CPCs at Frenchman's Reef in the Virgin Islands.

discussed in the previous section by shaping the outer glass envelope of the evacuated tube into the concentrator profile (Garrison, 1977; Ford, 1977). Improved performance results directly from integrating the reflecting surface and vacuum enclosure into a single unit. This concept is the basis of a new evacuated CPC collector tube which has a substantially higher optical efficiency and a significantly slower rate of exposure induced degradation than external reflector versions. These performance gains are a consequence of two obvious advantages of the integrated design:

1. Placing the reflecting surface in vacuum eliminates degradation of the mirror's reflectance and thus permits high-quality (silver or aluminum with reflectances $\rho = 0.91$–0.96) first-surface mirrors to be used instead of anodized aluminum sheet metal or thin film reflectors ($\rho = 0.80$–0.85) typical of the external reflector designs.

Fig. 10.16 Part of a 12,000-ft^2 evacuated tube CPC array deployed on the Illinois Department of Agriculture Building in Springfield.

2. The transparent part of the glass vacuum enclosure also functions as an entrance window and thus eliminates the need for an external cover glazing. This increases the initial optical efficiency by a factor of $1/\tau$, where typical transmittances $\tau = 0.88$–0.92.

The solar energy group at the University of Chicago has developed this concept (Snail, 1983) in collaboration with the staff of the GE Research Laboratory which fabricated the tubes. Eighty prototype tubes were built and 45 of these assembled into a panel with $\approx 2\ \text{m}^2$ net collecting area. The integrated stationary evacuated concentrator (ISEC) collector shown in Fig. 10.17 is an extended cusp tube CPC (Winston, 1978) matched to a circular absorber of diameter 9.5 mm with an absorber–reflector gap of 1.9 mm introduced for thermal isolation and mechanical clearance. The design acceptance half-angle

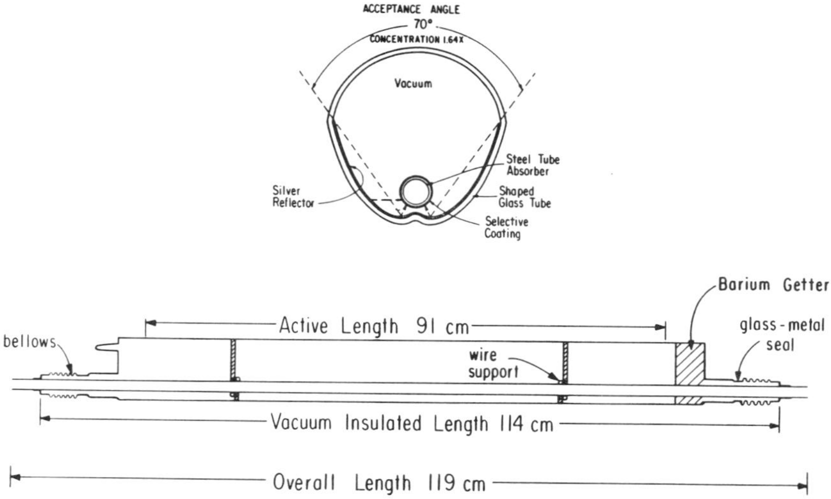

Fig. 10.17 Cross-sectional and longitudinal (not to scale) views of the integrated stationary evacuated concentrator (ISEC) design.

of $\theta = \pm 35°$ was chosen to permit stationary operation throughout the year. The maximum geometric concentration for this acceptance angle and gap is 1.85×; however, after truncating the CPC the net concentration was 1.64×. Special features of the mechanical design include: (1) double-ended unidirectional flow-through configuration; (2) copper glass-to-metal "housekeeper" seals; (3) metal bellows to allow for differential expansion between the absorber and the glass tube; and (4) a barium getter to preserve long lifetimes for hard vacuum ($<10^{-6}$ torr).

Despite the complications introduced by the flow-through metal absorber tube, this configuration was ultimately preferred because it provides the best heat transfer and the lowest pressure drops for a given size aperture. Because of the length taken up by the bellows, the seals, and the getter, the active length in the prototype was limited to 910 mm (36 in) and the relative effect of end losses was more important in the prototype panel than will be the case for a length optimized for production [probably ≈2 m (6 ft)].

In Fig. 10.18 is shown the transverse incident angle modifier, that is, the optical efficiency as a function of solar incidence angles in the

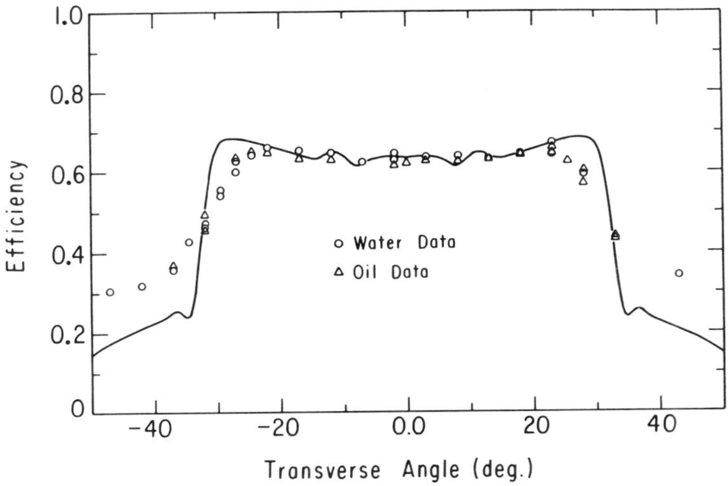

Fig. 10.18 Prototype ISEC efficiency versus transverse angle. The solid line is a cubic spline interpolation to results from a ray trace program run at 1° intervals on a single tube.

plane perpendicular to the longitudinal axis of the tubes. For comparison, the results of a ray-trace calculation for a perfectly aligned tube are shown. The average value of the efficiency across the flat top of the curve is $\langle \eta_0(\pm\theta)\rangle = 0.65 \pm 0.01$ and is in good agreement with the ray trace results out to ±25°. Both the measurements and ray-trace results show "shoulders" in the incidence angle modifier near the edge of the acceptance region around ±32° and a valley in the middle near normal incidence. This is due to the effect of "gap losses" (McIntire and Winston, 1981). The deviation of the panel data from the ray-trace results can be attributed to two effects: misalignment of absorbers within individual tubes and misalignment of tubes within the panel. This slight shortfall in angular acceptance (±32° versus ±35°) can be recovered by a slight redesign of the profile shape in the production version (Winston, 1980) and in any case results in only a <3% loss in annual energy delivery as calculated from hour-by-hour simulations.

The longitudinal incidence angle modifier for the ISEC panel is shown in Fig. 10.19. It is expected that the cosine corrected optical throughout will fall off slowly with increasing angle due to reflection losses at the glass and absorber surfaces, shading losses, and other

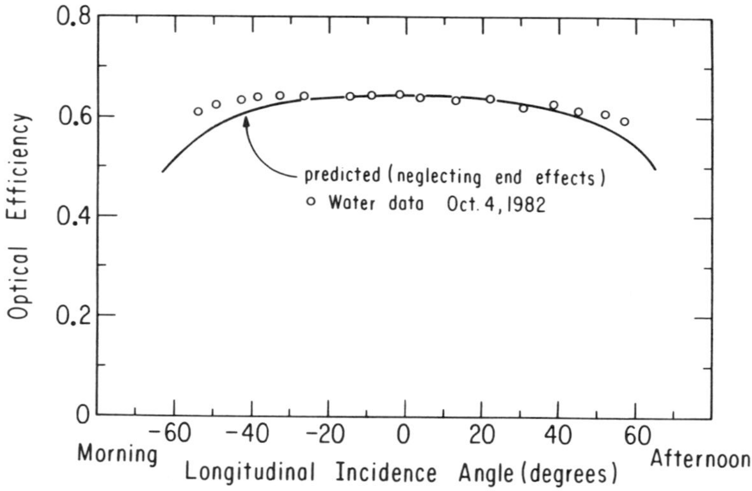

Fig. 10.19 Prototype ISEC efficiency versus longitudinal angle. The solid curve is a polynomial fit to data from a typical east–west collector.

similar effects. There is very little measurable effect ($<2\%$) out to $\theta \approx 40°$ and then a gradual drop-off at larger angles. Shown for comparison is a fourth-order polynomial suggested by Rabl (1985) which was derived from a fit to data for flat-plate and line focus troughs. The ISEC seems to be fairly well described by the same function.

Values of the solar thermal conversion efficiency measured at normal incidence and calculated according to Eq. 10.9 with $\eta_0 = 0.65$ $U = 0.2\ \mathrm{W/m^2K}$ and $U = 0.088$, including a correction for nonilluminated absorber length, are plotted in Fig. 10.20. The cluster of points near $(T_C = T_A)/I(\pm\theta) = 0$ represents the values of $\eta_0(\pm\theta)$ determined from measurements using water as the heat transfer fluid while those points at elevated temperatures used a high-temperature heat transfer oil. It should be noted that these data have been obtained with prototype tubes of length 0.91 m (3 ft) for which the relative importance of manifold and end losses is somewhat higher than the case of a practical production version of 1.8–2.4 m (6 to 8-ft) tubes. The good agreement between the experimental data and our predictions suggests that design improvements such as longer tubes, better selective coatings, and antireflected glass can all be treated with our simple model.

The performance data presented in the previous section consists mainly of instantaneous efficiency measurements made under clear

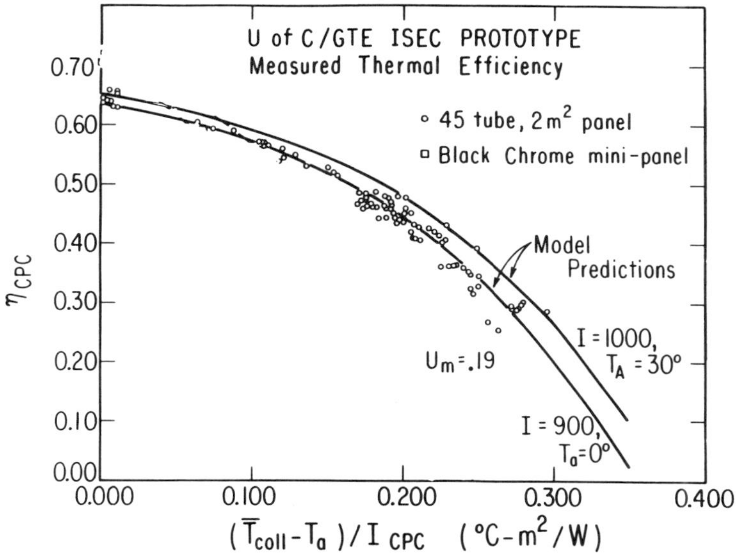

Fig. 10.20 ISEC thermal efficiency versus $\Delta T/I$ (CPC).

sky conditions at normal incidence, and as such, is not adequate for performing an economic analysis or comparing the ISEC to other collector types. For these tasks, we have developed a computer simulation program that integrates a collector's response to the solar input over a full year. The input weather data consists of a "typical meteorological year" (TMY) for 26 U.S. locations, obtained from the U.S. Department of Commerce's National Climatic Center. The simulations have been carried out for a variety of collector types, locations, and temperatures, and have been discussed in detail elsewhere.

In carrying out the simulations, a number of simplifying assumptions have been made:

1. The performance of the collector array alone is analyzed independent of the rest of the system; that is, any losses due to interaction with storage and load are neglected.
2. The calculations are carried out under steady-state conditions at each temperature (transient effects are assumed to be small).
3. Thermal losses are calculated as if the entire array were at one average temperature throughout the year.

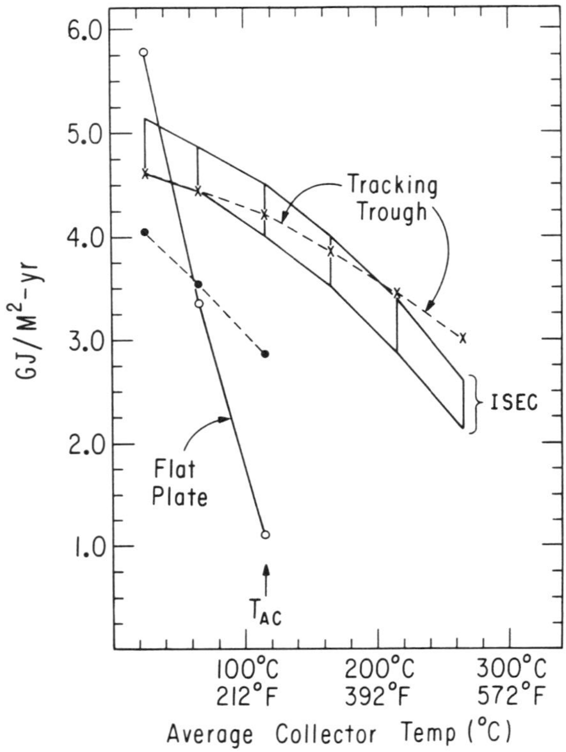

Fig. 10.21 Representative yearly performance versus temperature for four collector types in Phoenix, Arizona, latitude 33.4°N. The solid circles are for the contemporary evacuated tube collector.

The average yearly energy delivery in GJ/m^2 of net aperture is shown versus temperature for the ISEC and three other collector types in Fig. 10.21. In climates with a higher diffuse-to-total ratio such as Chicago and Boston, the ISEC will perform better relative to the parabolic trough. For reference, the instantaneous efficiency curves normalized to beam insolation for the four collectors are also shown in Fig. 10.22. The broad band labeled ISEC represents the range of performance expected from the present coating to a high-absorbance coating with $\alpha \approx 0.94$. The only other change in the prototype design that we have assumed is an increase in the active length from 0.9 to 1.8 m (3 to 6 ft). Note that, as in the instantaneous performance curves,

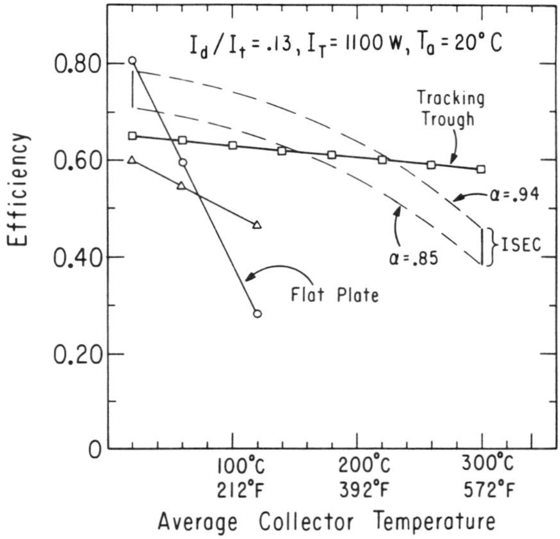

Fig. 10.22 Instantaneous performance curves relative to beam used in the long-term energy delivery calculations. The open triangles are for a contemporary evacuated tube collector with external reflector CPC and a dewar receiver.

the annual output of the ISEC collector is competitive with tracking troughs and dramatically improved over contemporary evacuated tubes and flat plates.

Over $1\frac{1}{2}$ years of operating experience with a 2-m^2 panel of prototype ISEC tubes has confirmed that there are no inherent technical problems with the design and that the measured performance is well approximated by our optical and thermal models. The design is very reliable. Although the tubes have been cleaned from time to time to remove ordinary dust and dirt, the underlying optical efficiency has remained unchanged over the entire period. Furthermore, only two tubes out of 80 produced have spontaneously lost vacuum during this time. This high level of reliability in a prototype collector was unexpected and should encourage ultimate widespread use of similar designs.

Two areas requiring further study are stagnation protection and alignment tolerances. The simplest solution to the first problem is the use of a selective coating that is stable to temperatures of ~450–500°C. Several candidate coatings are being investigated and all of them have the potential for a significantly higher absorbance ($\alpha > 0.90$) than the current coating. As an interim solution, geometric concentrations are being considered that would lower the stagnation temperature to ~400°C. The required alignment tolerances can be relaxed by oversizing the absorber slightly and alternate support structures, which would locate the absorber positively with respect to the glass cusp, should minimize misalignment effects.

Other design improvements being considered are:

1. Longer 2.1 m (~7 ft) tubes with the associated lower heat losses per unit net aperture
2. Modifications to the CPC profile to minimize gap losses and compensate for refraction through the cover
3. Antireflection treatment of the glass
4. A flattened cover to minimize glass volume, evacuation costs, and reflection losses

It should be emphasized that even without any of these improvements, the stationary prototype ISEC panel has exhibited performance comparable to commercial tracking parabolic troughs. Furthermore, with readily available materials and alterations, performance comparable to the best experimental troughs is expected to be achieved. No technical barriers remain and the obvious advantages of attaining this level of solar thermal collection efficiency from a simply nontracking collector should justify the additional development necessary for large-scale production and deployment of these collectors.

10.8 Nonimaging Secondary Concentrators

We have seen that the principles of nonimaging optics permit the design of reflector shapes of the CPC family which attain the maximum allowable geometric concentration for a given angular field of view. It is not so widely recognized that the same principles can be used to design secondary elements (Rabi and Winston, 1976) which can augment the concentration of more conventional focusing elements used as primaries so that the combined optical system can also approach the allowable limit. Applications of such two-stage designs lie in the higher-concentration small-angular-acceptance regime where the geometry of a single-stage CPC becomes impractical. The fundamental advantage is the same as in lower-concentration applications and may be expressed in complementary ways: either significant additional system concentration can be attained (i.e., a smaller lower-cost absorber) or the angular tolerances and precision can be relaxed while maintaining the same level of concentration. Two-dimensional (line-focus) configurations have been developed experimentally for both thermal and photovoltaic applications. Three dimensional (point-focus) geometries have been studied largely in the context of high-temperature thermal applications where the primary is a paraboloidal dish.

10.8.1 Intermediate Concentration for Photovoltaic Applications

A nonimaging secondary optical element can be used in conjunction with a conventional focusing lens, either to increase greatly the achievable concentration or to relax the required optical tolerances. At the present time, the most developed optical systems are Fresnel lenses. However, such systems require accurate tracking and alignment, typically to better than ±0.5°. In addition, it is very difficult to achieve concentrations over 500× with a Fresnel lens alone. Two-stage nonimaging concentrators employing Fresnel lenses as the primaries and dielectric totally internally reflecting concentrators as the secondaries offer a means of overcoming these difficulties. A two-stage design can deliver a desired concentration in the high-concentration regime (500–1000×), and in a more moderate concentration regime (80–100×), it can provide the maximum possible acceptance angle (Ning *et al.* 1987a,b). In photovoltaic applications, these concentrators provide much larger acceptance angles than conventional lens–cell systems with the same geometrical concentrations. It has also been shown that a two-stage concentrator provides a more uniform and consistent flux distribution on the cell surface and, therefore, the cell's electrical performance can be enhanced.

To demonstrate the technical feasibility and establish the performance of this approach, a prototype of such a two-stage photovoltaic concentrator (four-cell array) was designed, constructed, and tested. For this experiment the prototype is designed to achieve a concentration of 144×. The main advantage is seen as its large angular acceptance. Figure 10.23 shows the secondary assembly, while Fig. 10.24 shows the dramatic increase in angular acceptance from ±0.5° to nearly ±4°. In addition, we found that the optical losses in the secondary

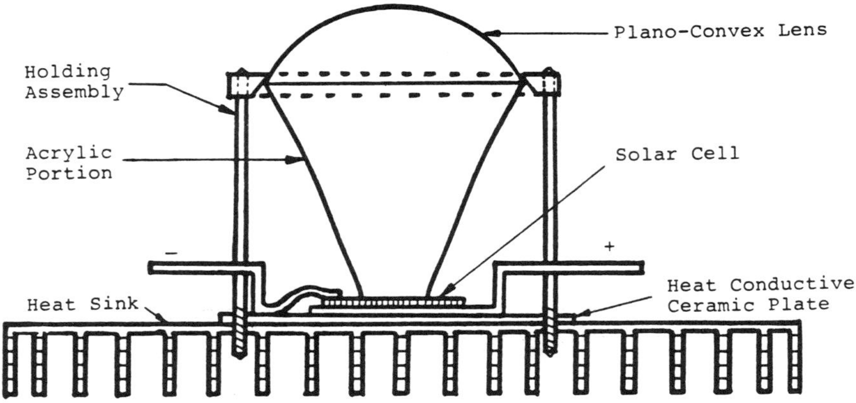

Fig. 10.23 Structure of the secondary-cell assembly.

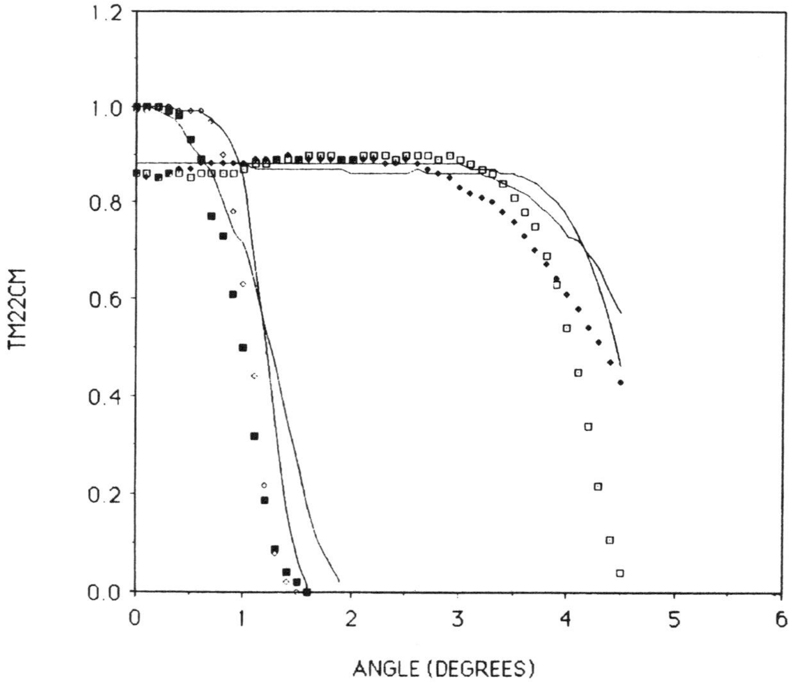

Fig. 10.24 Angular acceptance curves of the two-stage and the lens–cell systems. Their ray-tracing predictions are shown in solid lines.

(~15%) were essentially offset by improved electrical performance of the more uniformly illuminated cell. Thus, the on-track electrical performance of the lens or lens-with-secondary was the same, but the angular tolerance was increased by a factor of nearly 8.

10.8.2 High Concentration in Point-Focus Dish Solar Thermal Applications

The standard concave paraboloidal mirror is quite attractive for very large-scale concentration with a guiding system, and it can be combined with either a CPC-type or FLC nonimaging second stage to enhance the concentration. The paraboloid on its own has large coma and this means that at short focal ratios the gain in concentration is offset by the coma; for an input angle of ±0.25°, i.e., the sun, it is not worth going beyond about $f/0.5$, i.e., a numerical aperture of 0.7. Figure 10.25 shows how a compound elliptical concentrator (Section 5.4) can be added as a second stage; the compound elliptical concentrator (CEC) takes the aperture of the paraboloid as its source and its entry aperture is set at the focal plane of the paraboloid (Winston and Welford, 1980).

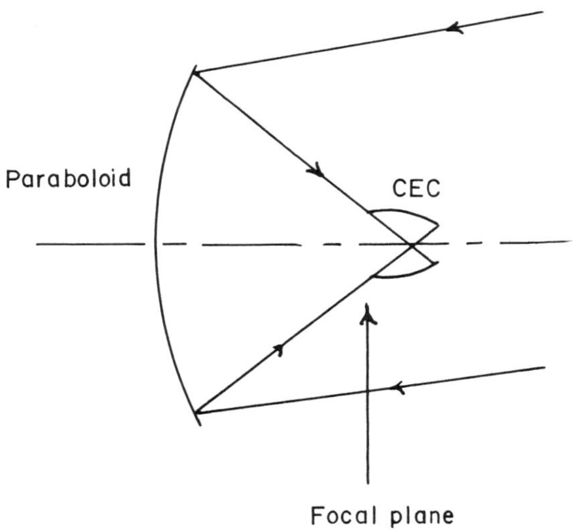

Fig. 10.25 A compound elliptical concentrator as second stage to a paraboloidal mirror; the mirror aperture is taken as the source for the CEC.

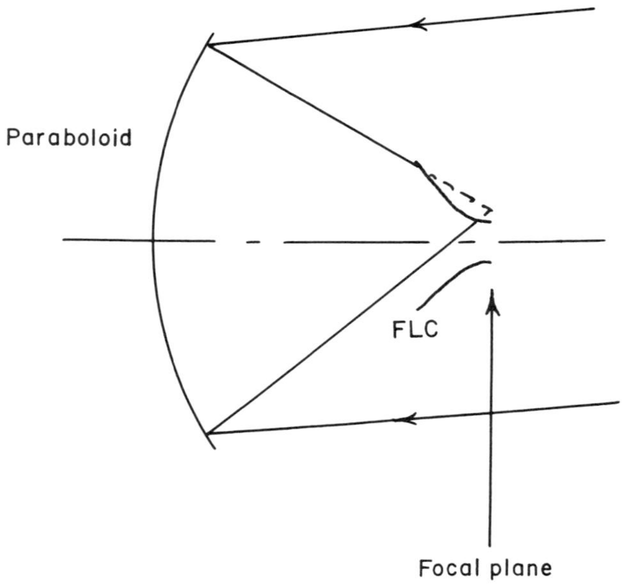

Fig. 10.26 A flow-line concentrator as second stage for a paraboloidal mirror. The virtual entry aperture of the FLC is at the focal plane of the paraboloid.

An alternative, which is in some ways more attractive, is to use a flow-line concentrator as the second stage, as in Fig. 10.26; in this arrangement the focal plane of the paraboloid coincides with the virtual entry aperture of the FLC (O'Gallagher and Winston, 1986).

Nonimaging secondaries such as the well-known CPC, deployed in the focal zone of image-forming primary concentrators, offer the optical designer an additional degree of freedom unavailable with any conventional approach. For a given angular acceptance, such two-stage configurations have the capability of delivering an additional factor of 2–4 in system concentration. In general, this is accomplished without doing anything to the primary. This unique capability can be used to increase either concentration or angular acceptance. For solar thermal applications these parameters are related, respectively, to increased performance (thermal efficiency) or relaxed optical tolerances (mirror slope error, angular tracking, etc.). Several alternative configurations are possible, including the CEC and the flow-line or "trumpet" concentrator in which the reflector is a hyperboloid of

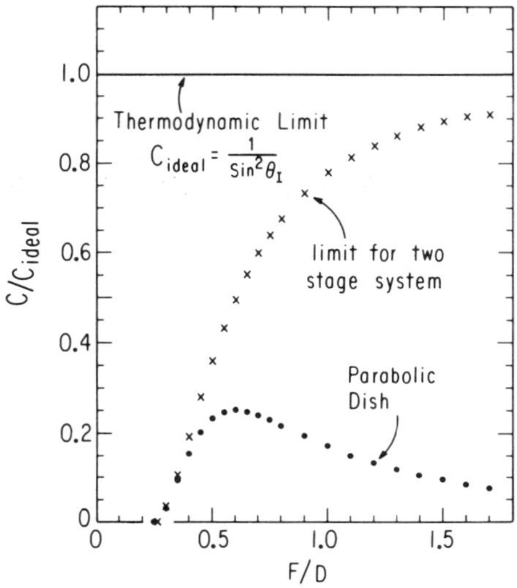

Fig. 10.27 The maximum geometric concentration for a given angular tolerance budget ($\pm\theta_I$) for one-and two-stage systems as a function of primary focal ratio. The two-stage configuration comes much closer to the thermodynamic limit, particularly at longer focal ratios.

revolution. The latter has been developed extensively, both analytically and experimentally. In addition, several versions of a secondary $5\times$ CPC were fabricated and tested at the Advanced Component Test Facility at Georgia Technical Institute (O'Gallagher *et al.*, 1987). Figure 10.27 illustrates the performance improvement provided by a nonimaging trumpet secondary as a function of design temperature, while Fig. 10.28 shows thermal performance, with and without secondaries in the "high performance limit" of very small slope errors. It is clear that in essentially all regimes of design parameters, the use of a secondary benefits performance.

10.8.3 Ultra-High Concentration of Solar Flux

As we have emphasized, the concentration limit of terrestrial sunlight is $1/\sin^2\theta_{sun} \approx 46{,}000$ in air, and $n^2 \times 46{,}000$ in a medium with index of refraction $n > 1$. However, imaging reflectors achieve less than 25% of this limit. By using a nonimaging secondary concentrator

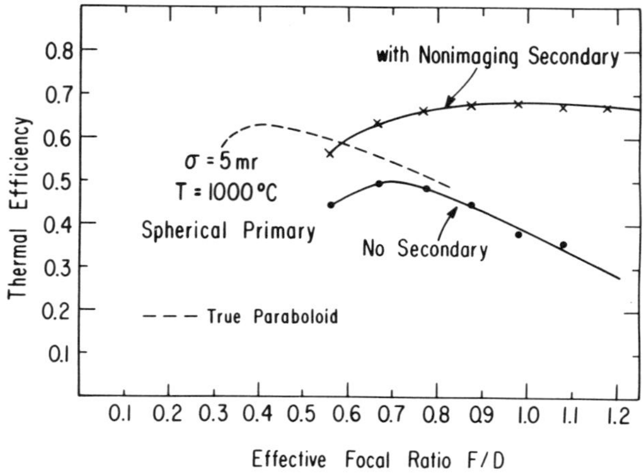

Fig. 10.28 Improvement in performance provided by combining a nonimaging secondary with a primary reflector.

in the focal plane of the primary mirror, the geometrical concentration can be increased by an order of magnitude. For example, the concentration factor for a primary alone is (neglecting shading)

$$C_{\text{primary}} \approx \cos^2 \phi \sin^2 \phi / \sin^2 \theta_{\text{sun}} \tag{10.10}$$

where ϕ is the rim angle. However, with the addition of a secondary element which is matched to the primary,

$$C_{\text{secondary}} \approx n^2 / \sin^2 \phi \tag{10.11}$$

So that the net concentration of the two-stage system is

$$C \approx n^2 \cos^2 \phi / \sin^2 \theta_{\text{sun}} \tag{10.12}$$

which could, in fact, be significantly higher than the limit in air of 46,000 even after various losses are taken into account. We have built a two-stage design that has a geometrical concentration of 102,000 by using an 11.5° rim angle primary, and a secondary of index 1.53 (Fig. 10.29). With this design one might expect to approach the irradiance that exists on the surface of the sun itself (7.3 kW/cm^2). A concentrated flux of 56,000 suns has been already achieved (Gleckman, 1988; Gleckman *et al.*, 1989). With such an intense light source, it should be possible to achieve efficient lasing in materials which efficiently absorb solar radiation and which may be tunable.

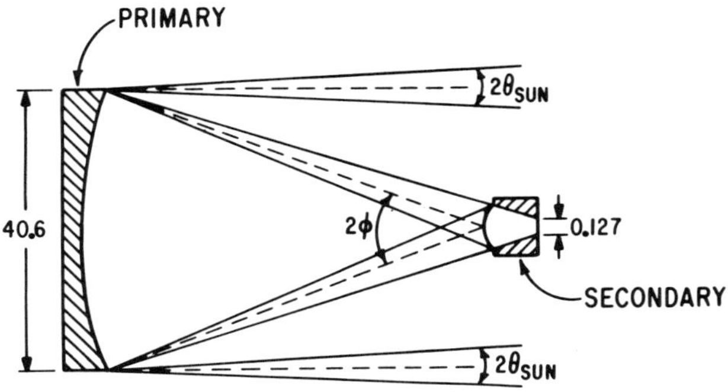

Fig. 10.29 Two-stage concentrator with refracting nonimaging secondary. The primary diameter and secondary exit diameter are each given in centimeters. Sunlight, incident from the right, reflects off the primary mirror and is reconcentrated by the secondary.

In all these applications, the methods and techniques of nonimaging optics permit us to concentrate solar flux at wide collection angles. This high collection optics can be used to advantage at every level of concentration in order to increase flux or, alternatively, to decrease requirements on tracking and on component tolerances and precision.

CHAPTER 11

Illumination of Optical Systems and Instruments

In this chapter we discuss applications of nonimaging optics to optical system design and to instrument design, with particular reference to utilization of light sources with maximum efficiency.

11.1 Introduction: The Radiance Theorem

Figure 11.1 shows a plane Lambertian light source of radiance B from which light is collected by a lens (condenser) of numerical aperture α_o. The lens therefore collects a total flux $\pi B \sin^2 \alpha_o$ per unit area from the source. If the source is then imaged onto an object plane with magnification m, the convergence angle will be α with $\sin \alpha = (1/m) \sin \alpha_o$, so that the irradiance (flux per unit projected area per unit solid angle) on the object plane will be B, neglecting transmission and reflection losses (but this irradiance occurs only within the solid angle of the cone of semiangle α). This result is a version of what is sometimes called the *radiance theorem*, or *brightness theorem*, according to which an image cannot be brighter than the original object, assuming the refractive indices in the source and object space are identical. The brightness theorem is, of course, included in the much broader Liouville's theorem as a special case. The relevance to the

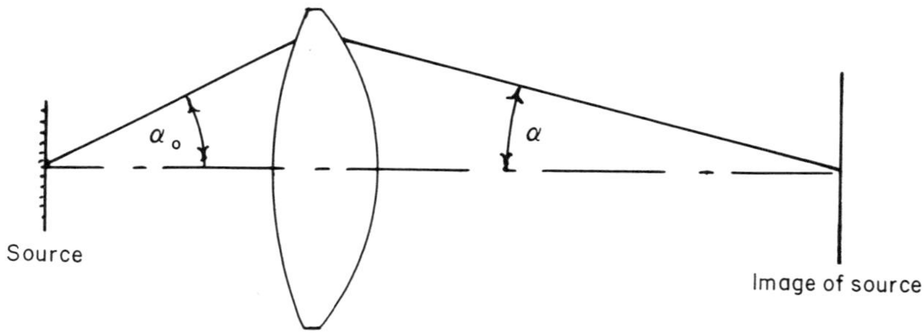

Fig. 11.1 A condenser lens and a plane Lambertian source.

present chapter is that in all optical systems, whether illuminated by a laboratory source or by some natural source, e.g., a star or a nebula, one could usually do with more light, and then if the source is Lambertian or approximately Lambertian, it is the radiance of the source, *not* its total flux output, which is usually the key. (Lasers are effectively point sources if they are single transverse mode and we discuss those separately in Section 11.4. Also if a Lambertian source is very small, this may be a restriction). In this chapter we shall discuss the illumination problem for imaging systems such as microscopes, projectors, etc., in which an extended incoherent source such as a Lambertian source can be assumed. Also we shall discuss the problems of using single- and multi-transverse mode lasers as illumination sources. The situations in which nonimaging optical systems could offer advantages will be noted.

11.2 Radiance of Laboratory Light Sources

The following table gives the radiance of a blackbody source at two wavelengths in the visible spectrum. The units are watts per micrometer wavelength interval per square millimeter per steradian.

	Wavelength (nm)	
Temperature (°K)	550	630
3000	0.39	0.60
3200	0.65	0.96
6000	30.7	27.3

A smooth tungsten surface has an emissivity of about 0.5, so the figures for 3000°K and 3200°K should be halved to get values for tungsten-halogen lamps run at these respective temperatures. The values for 6000°K correspond approximately to sunlight. High-pressure mercury and xenon arcs have a continuum spectrum with broadened spectral lines superimposed; the radiance of the continuum is approximately twice that of tungsten at 3200°K, and at the spectrum lines it is about another factor of 10 greater again.

Light-emitting diodes (LEDs) are becoming of increasing importance as small efficient sources in the visible and near-infrared regions. The light-emitting portion is a plane surface about 0.25 mm square, with a small portion of its area obscured by an electrode. The radiance of visible light LEDs is comparable with that of a tungsten lamp, allowing for the narrow spectral width of the light, but in the near infrared (say, 800–1000 nm) the radiance can be ten times greater than that of a lamp.

11.3 Conventional Illumination Systems for Instruments

The illumination systems for microscopes form a good illustration of the subject of this chapter. Figure 11.2 shows the simplest form of microscope illumination for transmissive objects. The condenser forms an image of the source on the object and this is viewed by the objective; the objective then projects an enlarged image perhaps onto a TV tube, a piece of photographic film, or even occasionally through an eyepiece

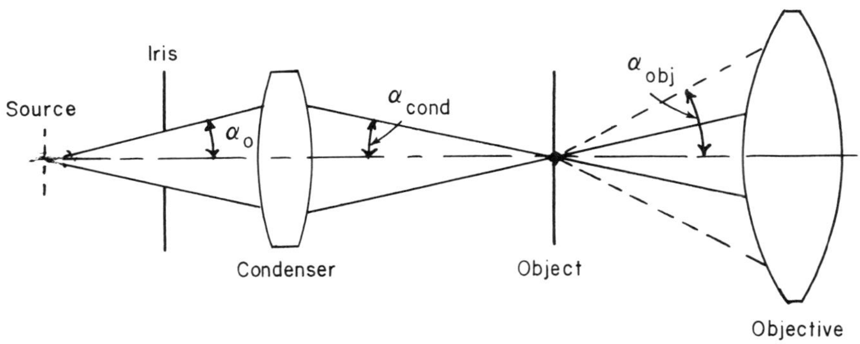

Fig. 11.2 Microscope illumination system.

onto the retina of an eye. The requirement is to illuminate a sufficient area with adequate angular subtense α, generally somewhat less than the angle α' collected by the objective. The point of the remark in the previous section about very small sources can be seen in this figure, for if the source is too small, it has to be magnified to cover the object and this may mean that the illuminating angle α is too small for the objective; conversely, if α is made large enough, the source image is too small. Clearly, one makes the best of this by collecting as large an angle α_o from the source as possible, but here there are practical limitations on conventional lens systems and it is not possible to have α_o much greater than 30° when taking light from a filament lamp.

A more elaborate microscope illumination system, the Köhler system, is shown in Fig. 11.3; as explained in the legend, in this system the iris diaphragms permit independent control over the illuminating angle and the area illuminated; also, the source image is not superimposed on the object. However, the limitation of too small a source is the same as for the simpler system in Fig. 11.2 and it is desirable to have as large a collecting angle at the first lens as possible for maximum light throughput. Of course, it should be said that there is often no shortage of light for ordinary visual use of a microscope; the more elaborate systems of illumination are intended for such purposes as television observation, photography, etc., where high speed may, for whatever reason, be important.

Fig. 11.3 The Köhler illumination system for microscopes. The substage condenser forms an image of the field iris at the object plane and the field lens forms an image of the source at the front focal plane of the substage condenser.

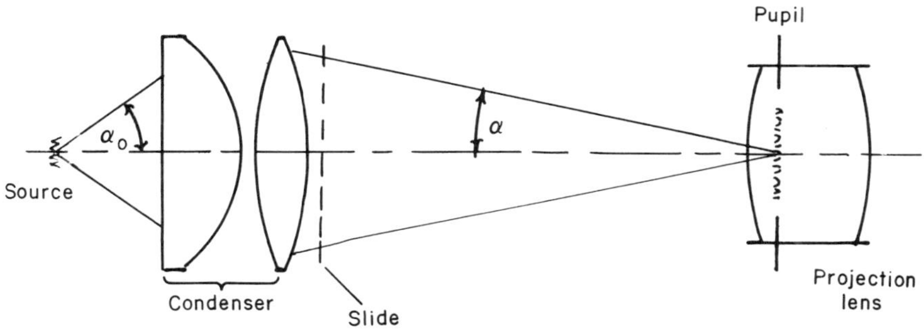

Fig. 11.4 Illumination for a slide projector; the condenser forms an image of the source in the entrance pupil of the projection lens; normally this image does *not* fill the pupil.

Figure 11.4 shows the condenser system of a slide projector; the double condenser, of which the first component has a nonspherical surface, produces an image of the source in the entrance pupil of the projection lens; the slide is immediately behind the condenser except, perhaps, for a heat-absorbing filter (not shown) and it is uniformly illuminated. In order to get a crisp high-contrast image, it is found to be best to underfill the pupil of the projection lens. The angle α is determined by the size of the slide and the focal length of the projection lens and thus α_0 is determined by the magnification needed for a given source size. Thus again we are left with the radiance of the source as the remaining parameter governing the brightness of the image. A smaller source with the same radiance can be used if there is enough magnification, i.e., if α_0 can be made large enough; this is the reason for the use of Fresnel condenser lenses in viewgraph projectors ("overhead" projectors); a Fresnel lens used for this purpose might have α_0 about 50°, but it would be quite impracticable to have a glass lens of the same size and numerical aperture as a condenser.

11.4 Nonimaging Optics in Light Collection for Instruments

Nonimaging optical systems such as CPCs used in reverse have something to offer in optical systems since in principle they permit collection over ±90° and they give an emergent cone of chosen angle

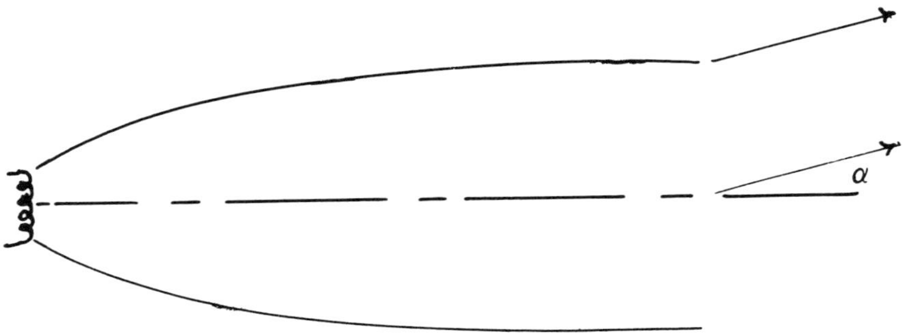

Fig. 11.5 A Lambertian source (filament) at the output aperture of a CPC; the emergent beam then fills the input angle of the CPC.

and minimum area according to Liouville's theorem. Thus we could imagine the filament of a tungsten–halogen lamp at what is normally the exit aperture* of a CPC, as in Fig. 11.5, giving an emergent cone of semiangle α over the minimum possible area as dictated by Liouville's theorem and with collection solid angle 2π.

There are three possible drawbacks to such a use of nonimaging devices in condenser optics: (1) the tungsten–halogen lamp has an envelope and so we cannot fill the exit aperture with the filament, (2) the filament has structure and this will mean that the distribution in angle and position at the entry aperture of the CPC will not be uniform, and (3) the CPC is much longer than a conventional lens system. (We recall from Chapter 4 that the length of a CPC is $a'(C+1)\cot\theta_i$, where a' is the exit semiaperture, C is the concentration ratio, and θ_i is the entry angle.)

To deal with the problem of the envelope, we can use conventional imaging optics to reimage the filament into the exit aperture. It may be objected to that the optical system to do this is limited in collection angle, so that there would be no gain, but there is one system which is not limited in this way and is very well suited to our present purpose; this is the concave spherical mirror, which forms an aberration-free image of a point object placed at its center of curvature. Figure 11.6 shows a hemispherical mirror with a filament object in the plane of the center of curvature; an inverted image of equal size is formed and

* To avoid confusion, we shall continue to use entry and exit for, respectively, the large and small ends of a CPC, although in the present context the light is traveling in the reverse direction.

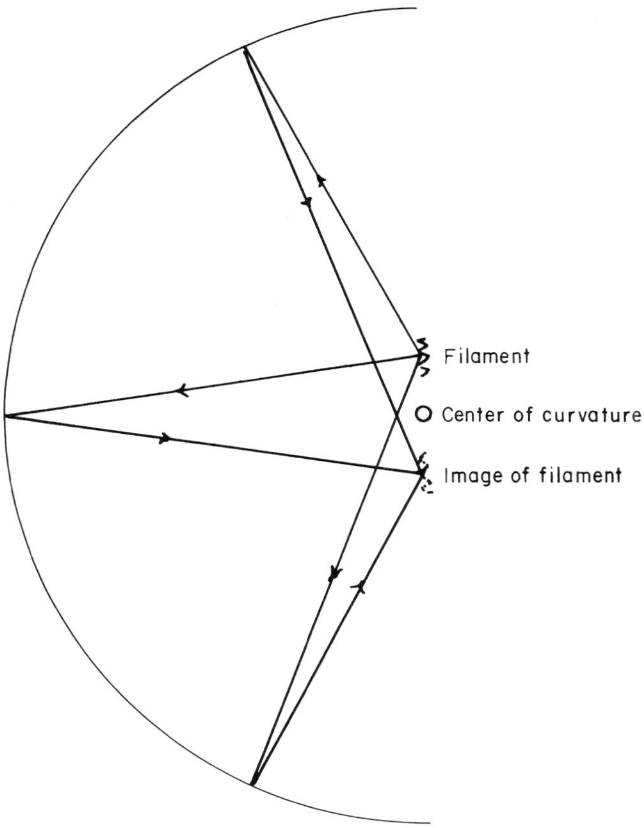

Fig. 11.6 Removing the envelope of a filament lamp. A concave spherical mirror is used to form a real image at unit magnification of the filament.

if the radius of curvature of the mirror is large compared to the size of the filament, the aberrations are small. The exit aperture of a CPC could then be placed at the filament image.

The structure in the filament image may or may not be a nuisance, depending on the application and on the design of the rest of the optical system. If it is a problem, the structure can be randomized by passing the light through a parallel-sided internally reflecting tube, sometimes called a *kaleidoscope*, as in Fig. 11.7. If the tube is long

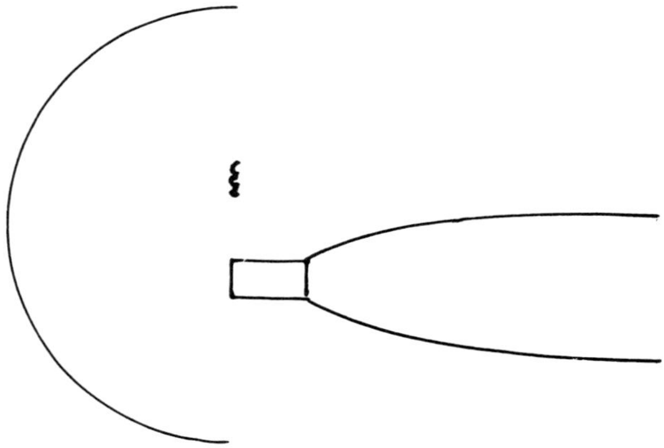

Fig. 11.7 A complete illumination system for a filament lamp; the glass envelope is removed by means of the concave mirror, the straight kaleidoscope section homogenizes the beam, and the CPC produces a uniform output within the desired solid angle. It is even possible to put another concave mirror behind the lamp to reimage the filament beside itself, so as to get, effectively, a larger source.

enough to ensure two or three reflections of most of the rays, the result should be adequate for any practical purpose.*

If the tube is solid, the refraction on entering ensures total internal reflection for all rays, provided the refractive index is greater than 1.414.

Many filaments have a roughly rectangular outline and, for maximum efficiency with these, the rectangular aperture CPC shown in Fig. B.3 of Appendix B would be advantageous; likewise the kaleidoscope would have a rectangular section.

An alternative to the kaleidoscope as a mixer is an integrating sphere with equal entry and exit apertures used at the real image of the filament where the solid angle of the light is 2π, as in Fig. 11.8; this would ensure very good mixing if the inside of the sphere is a good diffuser but it has the disadvantage that at least half the total flux is lost, since it returns through the entry aperture of the sphere.

There is little that can be done about the length of the CPC apart from some truncation (Chapter 5), but this may not be a serious disadvantage for many optical instrument purposes, since a reduction

* But it would, in theory, need an infinitely long kaleidoscope to produce perfect uniformity.

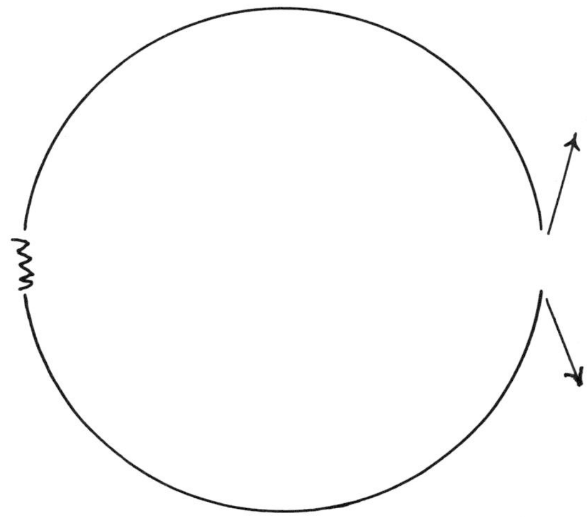

Fig. 11.8 An integrating sphere used to produce a Lambertian output; in principle, half of the flux reemerges from the entry aperture (if the apertures are equal in size) and, in practice, still more is lost by absorption at the nominally white diffusing coating of the sphere.

of the collection angle from 90° to, say, 30° would then permit the use of conventional lens optics for the rest of the illumination system and a CPC with concentration ratio 2 (in 2D) is very short.

The semiconductor surface of an LED emits over 2π steradians and its étendue is therefore 0.39 mm^2 sr; however, it is not always possible to get at this full étendue because of the encapsulation of the diode; this can take the form of a solid epoxy lens arranged to produce either a wide- or medium-angle beam, but these lenses are usually not accurately molded and the aberrations cause loss of étendue. It is also possible to have the LED unencapsulated and either protected by a plane window at a millimeter or so distance or left bare. The ideal would clearly be to incorporate a CPC in the encapsulation if maximum flux collection is desired.

11.5 Lasers as Sources for Optical Instruments

The ordinary single-mode cw laser, such as a standard 1 mW HeNe laser, is fundamentally different as a light source from thermal sources such as filament lamps and discharge tubes; it is effectively a single point source in that it produces a beam of light in which at any point there is only one phase front, rather than an infinity of phase fronts ranging over a certain solid angle corresponding to the angular subtense of the original thermal source. Thus the light from a single mode laser could be said to have zero étendue.*

The single transverse-mode laser is completely coherent across the beam, i.e., light from any two portions of the beam can interfere with good contrast. This leads to the formation of coherent noise or "speckle" in images formed in laser light: Thus if a transparency is illuminated by a laser, regions which have the same optical density but different phase retardation will produce interference effects, i.e., speckle, in the image. Much effort has been devoted to finding ways of despeckling images (see, e.g., McKechnie, 1975); all methods amount, in effect, to superimposing a number of images with different speckle. If N images with statistically independent speckle are superimposed, the contrast in the speckle is reduced by the factor $N^{\frac{1}{2}}$.

Figure 11.9 shows a transparent object illuminated by an expanded laser beam and imaged onto, say, a TV camera; then the simplest way to reduce the coherent noise is to introduce a ground glass diffusing screen as indicated, move it about between frames, and average the successive TV frames. This, of course, is very wasteful of light. A more efficient way is to illuminate the object obliquely, as in Fig. 11.10 and to vary the angle and/or azimuth of the oblique illumination between frames (or exposures if the object is being photographed).

It is *not* enough merely to introduce a ground glass diffuser to jumble the phases of the different parts of the beam, since all parts will still be coherent with each other and the screen will produce very high-contrast speckle; it is essential to move the screen to produce different phase patterns and then to superimpose them *incoherently* by addition of intensities. This means that it would not help to, say, send the laser beam in through the exit end of a CPC as a static device.

* However, this is not very helpful and if it were required to assign an étendue or phase space volume to a laser beam, it would be better to introduce the delta function formalism. Manufacturers of lasers customarily quote a "divergence angle," but this has nothing to do with finite source size; it refers to the rate of spreading of the beam due to diffraction and the finite curvature of the phase fronts. Thus for a laser beam of divergence angle ε and power P originating at (x_o, y_o) we could say that the étendue is $\pi P \varepsilon^2 \delta(x - x_o)\delta(y - y_o)$. In practice it is unlikely that such a formal device would be required for discussing laser illumination.

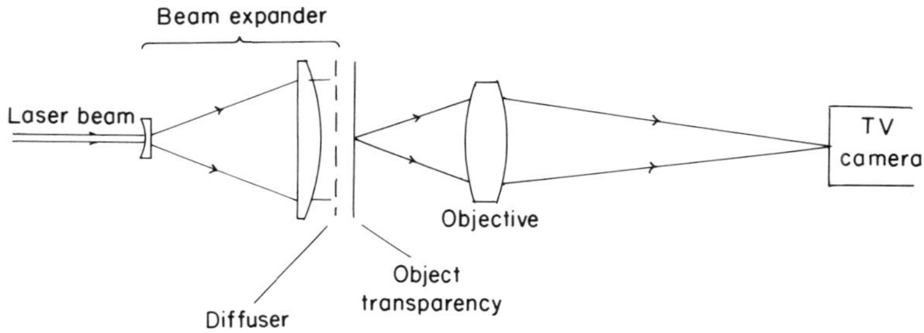

Fig. 11.9 Decohering a laser beam as an illumination source for an imaging system. The diffuser must be moved between frames, so that the system can accumulate an average of N images, which are all incoherent with each other; then the variance of the intensity of the speckle noise is reduced by the factor N.

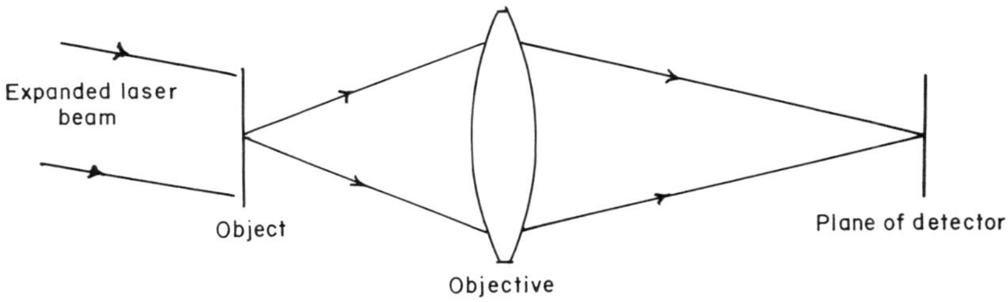

Fig. 11.10 Oblique illumination by an expanded laser beam; if the azimuthal direction of the illuminating beam is changed between frames and an average is taken, the effect will be to reduce speckle noise, as in Fig. 11.9, but without the loss of flux due to the diffuser.

However, an efficient way to use a diffusing screen with a laser and CPC is shown in Fig. 11.11; the diffuser fills the exit aperture and the CPC gives a uniform beam (apart from speckle) of chosen angular range; then, on moving the diffuser and averaging, the speckle contrast is reduced as before, but since the CPC collects over 2π solid angle, the only loss of flux is from that which is scattered backward by the diffuser.

HeNe lasers and many others have a long coherence length, i.e., interference can be obtained between beams with path differences of hundreds of millimeters, but some lasers, notably excimer lasers, which give wavelengths in the ultraviolet, have very short coherence lengths,

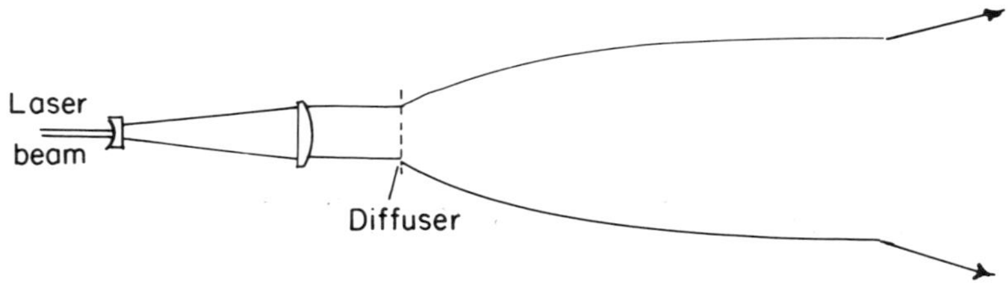

Fig. 11.11 Using a laser with a CPC. The expanded beam meets a diffuser and this provides, *in the geometrical optics model*, a Lambertian source at the output aperture of the CPC; the emergent beam is then distributed over the input angle of the CPC, but the input aperture is still a coherent source and it is necessary to move the diffuser and average over multiple exposures to avoid speckle.

usually a fraction of a millimeter. This gives the possibility of a method of despeckling which does not involve moving optical components. The principle is shown in Fig. 11.12. The expanded laser beam meets an array of lens prisms which expand and deflect the light so that the various beams meet at the object, arriving from different directions; the lens prisms are made of different thicknesses so that each beam arrives with a delay which, relative to any other, is greater than the coherence length and thus it cannot interfere with any other beam. The effect is then of a number of mutually incoherent beams equal to the number of lens prisms. Systems of this kind, known as *fly's-eye condensers*, have been in use for many years (but without deliberately introduced path differences) for illuminating projectors from very small intense sources such as high-pressure xenon lamps. Here again

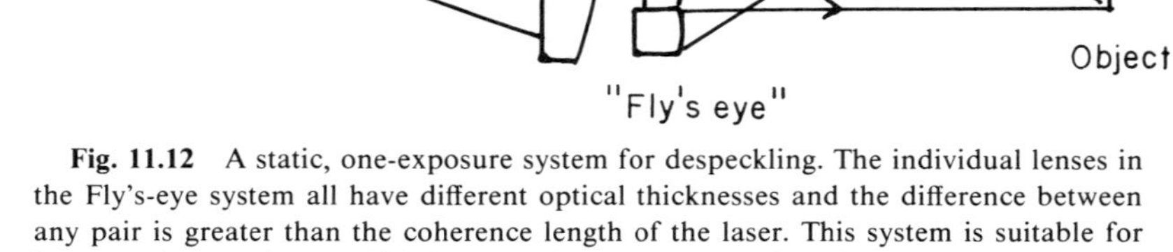

Fig. 11.12 A static, one-exposure system for despeckling. The individual lenses in the Fly's-eye system all have different optical thicknesses and the difference between any pair is greater than the coherence length of the laser. This system is suitable for lasers such as excimers which have a coherence length of a millimetre or less.

a nonimaging system could be used to improve uniformity, as a kaleidoscope, or to change angular subtense, as a CPC.

11.6 Fiber Optics Applications

Many medium- and low-power laser diodes are single transverse-mode devices (at any rate, when run at full power) but they emit an elliptical cone which might typically have semiangles of 10° and 30°, as in Fig. 11.13.* A multimode fiber for data transmission might have a numerical aperture of about 0.2, i.e., semiangle 12°. The diode laser can be efficiently coupled into the fiber by means of a nonimaging device which collects at up to 30° and which has an exit angle of 12°, i.e., a solid θ_i/θ_o concentrator used in the reverse mode from that described in Chapter 5. The exit from a diode laser is only a few micrometers across where it leaves the laser material; this makes it possible to have the concentrator small enough to ensure that its exit aperture matches the entry diameter of the fiber, typically 5–25 μm in diameter.

Nonimaging systems are a convenient alternative method of coupling multimode fibers end to end. The problem in coupling is to align accurately two fiber ends of very small diameter and ensure good optical contact. If each end is attached to a solid θ_i/θ_o concentrator,

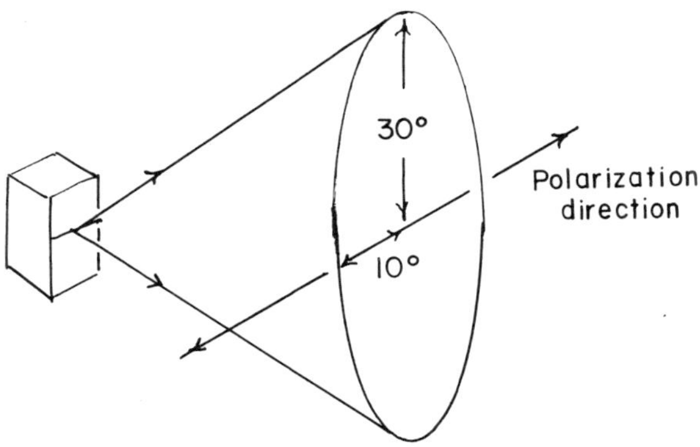

Fig. 11.13 The shape of the beam from a laser diode.

* In a recent development, laser diodes which emit a symmetric cone beam have become available.

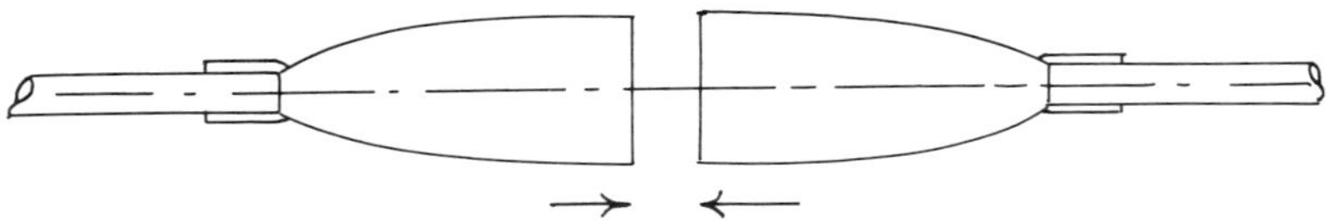

Fig. 11.14 Connecting two fibers by means of solid CPCs.

as in Fig. 11.14, with a small enough value of θ_i, then the two large-diameter ends of the concentrators can easily be aligned mechanically. The concentrators can be plastic moldings with a hollow section at the small end to guide the fiber.

Multimode fibers are used simply to guide light to convenient places in instruments, and LEDs form convenient light sources for this purpose. The étendue concept can be used to show how many fibers can be fed from one LED. Thus a typical multimode fiber would be 25 μm in diameter with numerical aperture 0.2, so that its étendue is 0.000062 mm^2 sr; but from Section 11.4 the étendue of an LED is about 0.39 mm^2 sr, so that in principle some 6000 fibers could be fed from one LED. In practice there would be considerable packing losses from the fiber cladding, etc., but still it would be possible to convert the LED exit from NA 1 (i.e., 2π sr) to NA 0.2 by means of a CPC and then connect directly to a fiber bundle containing perhaps 3000 fibers, as in Fig. 11.15.

11.7 Collection of Čerenkov and Other Radiation

The earliest application of nonimaging concentrators of CPC type was to the collection of Čerenkov radiation for detection by photomultipliers. The problem in such applications is to collect a relatively small light fiux from a volume of gas of perhaps several cubic meters and to detect the flux by means of photomultipliers that have photo-

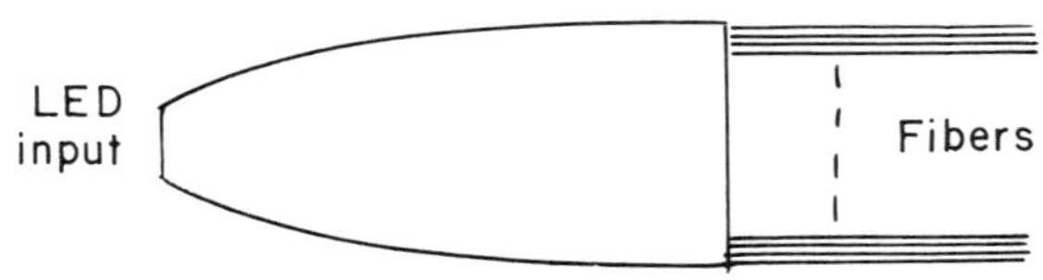

Fig. 11.15 Light emitting diode input to a CPC to feed a bundle of fibers.

cathodes of diameter at most about 120 mm. The light is emitted at angles less than a certain limiting angle θ_i measured from the direction of an incident particle beam; clearly, a case where CPCs of concentration ratio $1/\sin^2 \theta_i$ could be useful. This application was proposed by Hinterberger and Winston (1966a,b) and such systems have been constructed on a large scale at the University of Chicago and Argonne National Laboratory (Hinterberger *et al.*, 1970). In one system 28 CPCs of concentration ratio 4 were used to feed 28 photomultipliers with photocathodes about 110-mm diameter. Thus a fourfold economy was achieved in the number of photomultipliers and in the related power supplies and auxiliary electronics.

In another application, also calling for economy in photocathode area, Hinterberger and Winston (1968a) used a solid dielectric CPC coupled to Plexiglas light guides from a sheet scintillator. The same authors used a two-stage system of a $\theta_i\theta_o$ concentrator followed by a solid CPC (Fig. 5.8) to reduce the overall length of a concentrator (Hinterberger and Winston, 1968b). The solid CPC was put in optical contact with the end window of a photomultiplier in which the photocathode surface was deposited directly on the inside of the window. Thus an increase in efficiency by a factor n^2 was achieved, i.e., in effect a two to threefold reduction in the number of photomultipliers and associated electronics.

11.8 Concentration of Radiation in Detector Systems Limited by Detector Noise

Detectors for the infrared beyond a wavelength of a few micrometers are limited in performance by noise originating in the detector, in contrast to detectors such as photomultipliers in which the main source of noise is usually shot noise from the signal and background radiation. The detector noise, which is due to a combination of several causes, is conveniently measured by the *noise-equivalent power* (NEP), which is the signal power that would give a signal-to-noise ratio of unity. The NEP is generally proportional to the square root of the area of the detector. Thus in order to maximize the signal-to-noise ratio, it is desirable to use as small a detector as possible. In fact, most detectors for the infrared have active surface areas of order 1 mm^2 or less. Now consider the problem of using such a detector to scan the image from an F/16 telescope operating at a wavelength of, say, 500 μm. The point spread function or impulse function of the telescope has a central lobe of half-width about 10 mm and if we were to scan this with a 10-mm-diameter detector, the detector noise would be ten times that from a 1-mm detector.* Using conventional field optics systems such as off-axis mirrors or small infrared transmitting lenses (Smith, 1966), typically ~F/2, one can obtain a concentration of $\sim(2/0.5)^2 = 16$ with a corresponding reduction in the time required to achieve a given signal-to-noise ratio. That this is clearly a case for a nonimaging concentrator was first recognized by Hildebrand (Harper *et al.*, 1976). The University of Chicago group performed laboratory trials and observational tests of CPCs used to concentrate flux from a chosen aperture of the image plane to a diameter matching an infrared detector. Working independently, a University of California group (Richards and Woody, 1976) applied the technique to measuring the spectrum of the submillimeter cosmic background radiation.

The application of CPCs to bolometer detectors poses a special problem: Conservation of étendue necessarily implies that the radiation exits the concentrator over an entire hemisphere. On the other hand, bolometers are insensitive to radiation incident at large angles (the large index of refraction causes high reflection losses). Moreover, black coatings used to improve absorption become ineffective at long wavelengths. Following a proposal by Hildebrand, the University of Chicago group approached this problem by placing the bolometer inside a cavity. By suitable choice of the dimensions of the cavity and of the position of the bolometer within it, a uniform response over the entire angular acceptance of the CPC is achieved. The authors have named this combination of CPC concentrator with matched cavity a *heat trap.* For example, with the Hale 200-inch telescope used at prime focus (F/4), they used a CPC with a 17.5-mm-diameter entry aperture and 2.2-mm-diameter exit aperture. The detector, a cooled bolometer, was in a matched cavity leading off the exit aperture. At a wavelength of 1.4 mm the scan across Jupiter agreed well with what would be expected from a point source image. Since Jupiter is unresolved by the Hale telescope at this wavelength, this result was to be expected. In principle, heat trap field optics gives a fourfold improvement over conventional F/1 field optics for extended objects or for

* There would be other problems, such as increased time of response because of the greater heat capacity of the 10-mm system.

pointlike objects where the central lobe of the diffraction pattern is much greater than the entrance aperture. In practice, some improvement is observed even when these conditions are not satisfied because the nonimaging system has better efficiency for the peripheral rays. Comparison with other systems is difficult because of differences in bolometer loading, etc., but generally one obtains an improvement of ~2 for *point* objects just *below* the diffraction limit.

CPCs are now used in most broad-band instruments for far-infrared and submillimeter astronomy. A practical advantage of the CPC over off-axis mirrors, in addition to their better concentration, is that they are more easily incorporated into detector arrays (D. A. Harper and R. H. Hildebrand, in preparation) and into far-infrared polarimeters (Hildebrand, Dragovan, and Novak, 1984; Novak *et al.* 1989) where it is essential to have optical elements that are compact and easily aligned.

It should be noted that in applications to far infrared astronomy we are no longer in the region where the design performance can be assessed solely on the basis of geometrical optics. Operating near the diffraction limit, the exit aperture of the concentrator is ~1 wavelength in diameter and under these conditions we should expect diffraction effects to be very significant. It would be very difficult to calculate fully the performance of one of these concentrators, taking account of diffraction even as a scalar effect, and much more so to do a full treatment according to electromagnetic wave theory, since this would involve solving Maxwell's equations for boundary conditions defined by an obstacle many wavelengths in size and of a shape described by a relatively complicated equation (see Chapter 8 for some discussion of these problems). Actually, such solutions have been carried out only for rather simple shapes such as spheres, disks, ellipsoids, slits, etc. We can point out two ways in which diffraction would change the performance from the geometrical optics predictions. First, the entry aperture was some 12 wavelengths in diameter. Now the very crude approximation of Kirchhoff diffraction theory shows that some flux is directed into the concentrator at larger angles than the incident angle after passing the entry aperture and some of this might be expected to be rejected, thus decreasing the overall transmission. In fact, it is known that the Kirchhoff theory is very inaccurate for such small apertures, and more precise scalar or vector theories show that the transmission of a plane circular aperture may be either less than or greater than the geometrical optics prediction, depending on the ratio of the diameter to the wavelength and on what assumptions are made about the interaction of the material of the aperture with the radiation, e.g., whether reflecting or absorbing. The same argument then applies even more strongly to the exit aperture. However, such a small aperture surrounded by the strongly curved reflecting surface of a CPC is clearly a very different problem from a plane reflecting screen with a circular aperture. We can guess that the actual result could be very different from the geometrical optics prediction. In fact some preliminary experimental results by J. Keene (personal communication) on lens–mirror CPCs used in this way suggest that the transmission is appreciably higher than would be expected from ray tracing and Kirchhoff diffraction theory.

It seems that this application could be extended to a variety of infrared detection problems. Some of the most sensitive medium-wavelength ($\sim 10\,\mu m$) detectors are mercury–cadmium–telluride photovoltaic devices cooled to liquid nitrogen temperature, and (as currently available) the sensitive area is less than 1 mm^2. In order to get the most out of such detectors at low intensities, it could be advantageous in many cases to use a CPC, since the greater concentration ratio compared to image-forming systems would make it possible to collect radiation from a larger source, e.g., the exit grill of a spectrometer.

11.9 Stray-Radiation Shields

A different example of the possible uses of CPCs is for *excluding* unwanted radiation. In many astronomical applications, particularly for infrared work, it is important to exclude unwanted radiation coming from directions other than that of interest, and elaborate systems of baffle plates and diaphragms are used. Figure 11.16 shows how a CPC might be used for this purpose. For simplicity we consider a refracting telescope objective that is required to cover an angular

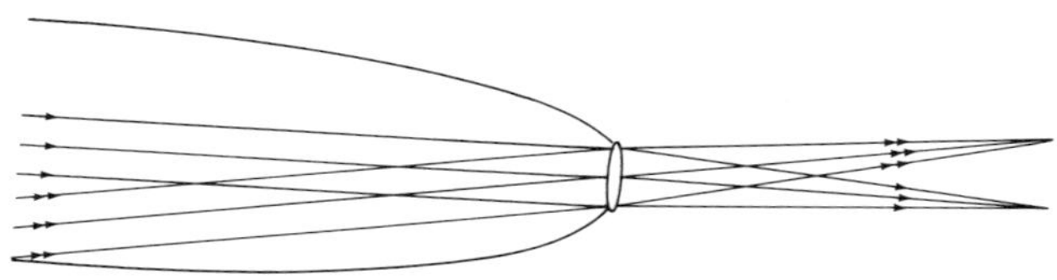

Fig. 11.16 A CPC used to reject light at unwanted angles from a telescope.

field of view 2β. Then it is easy to see that a CPC arranged as in the figure would *reflect* back all rays at angles greater than its acceptance angle θ_i. An approximate value for θ_i which would permit all rays at angles less than β to enter is

$$\theta_i \sim \beta(1+2\beta) \tag{11.1}$$

The useful aspect of such an arrangement would be that the unwanted rays are actually reflected back rather than absorbed on baffles, and thus heating of the telescopes and detectors is avoided.

A similar method could be used in reverse to shield signal lamps, so as to ensure that their light could only be seen over a previously specified range of angles.

11.10 Optical Pumping and General Condensing Problems

We noted in Section 6.5 the possibility of using a concentrator generated by the string construction to replace the customary cylindrical ellipsoidal cavity used for connecting a pumping flash tube with a laser rod. The advantage here would be gains in pumping power and in uniformity of illumination of the laser rod.

An allied problem is that of illuminating a solid light guide which may have an acceptance angle as high as ±90°. There is usually a considerable loss in illuminating these systems with a conventional lens system, for reasons explained in Chapter 3, unless the source has a very much greater surface area than the end of the guide. However, a nonimaging concentrator can be designed by the method in Chapter 6 that will send the theoretical maximum flux down the guide if the source has the same surface area. This is illustrated in Fig. 11.17.

As implied at the end of the last section, if a lamp is placed at the absorber end of a CPC, the light emerging from the entry end is very

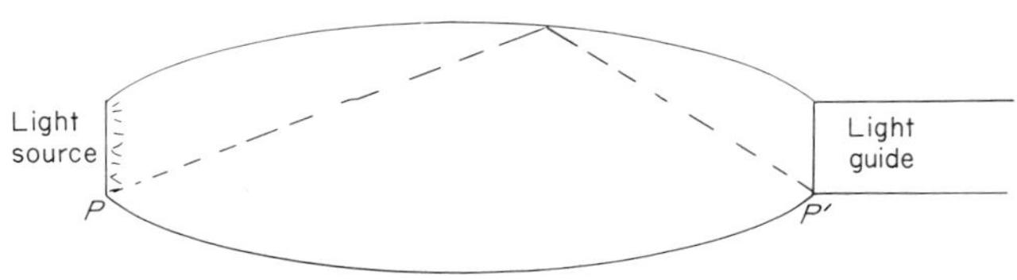

Fig. 11.17 A concentrator for illuminating a solid light guide. The shape is constructed by rotating the ellipse with foci at P P′ about the axis of symmetry.

sharply delimited in angular range. This has perhaps slightly frivolous applications in domestic, commercial, and stage lighting, where a sharply defined pool of light is required. Desk and table lamps designed on this principle are beginning to be manufactured.

11.11 Biological Analogs

The mode of action of the rod and cone receptors in eyes of many different kinds has long been a subject of debate, and it is certain that we do not yet know the full story. As a small contribution to this problem, it was noted by Winston and Enoch (1971) that the shape of the tapering portion of a human retinal cone is very similar to that of a CPC with entry angle $\theta_i \approx 13°$. The refractive index difference between the cone and the outside medium is quite small and this means that the transmission-angle curve is considerably modified from the ideal, as in Fig. 11.18. The drop in transmission with angle of incidence has a very suggestive resemblance to what occurs in the Stiles–Crawford effect, in which light passing through the rim of the pupil is less effective in vision than light passing centrally. However, this should not be taken too far, since other mechanisms, including a waveguide effect and some polarization effects, almost certainly are also involved in the cones.

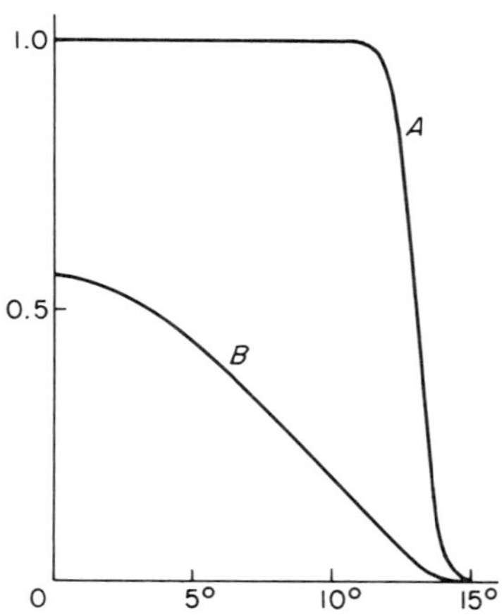

Fig. 11.18 The human retinal cone as a light guide. Curve A shows the transmission angle characteristic of a 13° CPC. Curve B is for a CPC of similar angle but with refractive index ratio between inside and outside 1.04 and with some allowance for absorption by visual purple. (After Winston and Enoch, 1971).

A similar resemblance in the cones of the crab *Limulus polyphemus* was noted by Levi-Setti *et al.* (1975). The cones of this animal form a compound eye (fly's eye) and they are about 0.2 mm in diameter, i.e., some hundred or so times larger than human cones. Thus one would expect theories based simply on geometrical optics to have considerable validity for this structure. The authors cited found that the profile of a *Limulus* cone was very closely approximated by a CPC with entry angle $\theta_i = 19°$. This would clearly be the most efficient shape for a single unit of a compound eye designed particularly for seeing in low illumination levels.

APPENDIX A

Derivation and Explanation of the Étendue Invariant, Including the Dynamical Analogy; Derivation of the Skew Invariant

A.1 The Generalized Étendue

In Section 2.7 we introduced the invariant

$$n^2\, dx\, dy\, dL\, dM \tag{A.1}$$

The meaning of this was stated to be as follows. Let any ray be traced through an optical system and let rectangular coordinate axes be set up in the entry and exit spaces in arbitrary orientations. Also, let the ray meet the x, y plane in the entry space in (x, y), and let its direction cosines be (L, M, N), and similarly for the exit space. Then for any nearby ray with coordinates $(x+dx, y+dy, L+dL, M+dM)$ we have

$$n'^2\, dx'\, dy'\, dL'\, dM' = n^2\, dx\, dy\, dL\, dM \tag{A.2}$$

We shall prove in Section A2 that this result is true for any optical system and for any choice of the directions of the axes. But first we discuss some different ways of interpreting this result. We put $p = nL$ and $q = nM$, and we treat (x, y, p, q) as coordinates in a four-dimensional space. Let U be any enclosed volume in this space. Then

U is given simply by

$$\int dU = \iiiint dx\, dy\, dp\, dq \tag{A.3}$$

taken over the volume. The result of Eq. (A.2) means that if we let a ray sweep out the boundary of this volume in the four-dimensional entry space, it will sweep out an equal volume in four-dimensional exit space, or, clearly, in any intermediate space of the optical system. There is a useful analogy between this space and the multidimensional phase space of Hamiltonian mechanics, and so we shall call it simply phase space. Thus our result is that if a ray sweeps out a certain volume of phase space in the entry region, it sweeps out the same total volume in the exit region, or, as it may conveniently be put, phase space volume is conserved. Of course, the shape of the volume will change from entry to exit space, or even if the origin of coordinates is shifted parallel to the z direction. However, it can easily be shown that rotating the axes about the origin does not change the phase space volume, so that it is a physically significant invariant.

Another picture that is particularly valuable for the purposes of this book may be obtained by supposing that an area in the x, y plane (in real space, not phase space) is covered with uniformly closely spaced points and that through each point a large number of rays is drawn. These rays are to be in different directions so that their direction cosines increase by uniform small increments over certain ranges of L and M. The rays so drawn then represent "ray points" uniformly spaced throughout a certain volume of phase space, and our theorem can then be stated in the following form: The density of rays in phase space is invariant through the optical system. If the four-volume U of Eq. (A.3) has been defined by means of physical components in the optical system, i.e., an aperture stop and a field stop, it is known as the *throughput* or *étendue* of the system.

We can see from the ray-point representation that the étendue is a measure of the power that can be transmitted through the optical system from a uniformly bright source of sufficient extent, in the geometrical optics approximation. Here, of course, in addition to neglecting interference and diffraction effects, we are assuming no losses in the system due to reflections, absorption, or scattering. The concept of étendue is applied in comparing the properties of many different kinds of optical instruments, although usually only a simplified version appropriate to axially symmetric systems is used.

A.2 Proof of the Generalized Étendue Theorem

Equation (A.2) has been proved in many different ways. In particular, the analogy from Hamiltonian mechanics can be used to give a simple proof (see, e.g., Winston, 1970; Marcuse, 1972). However, since we are here concerned with applications to geometrical optics, it seems appropriate to give a proof based directly on optical principles. We shall discuss the mechanical analogies in some detail in Section A.3.

We follow closely the method of Welford (1986), which makes use of the point-eikonal, or characteristic, function of Hamilton. This function is defined as follows. We take arbitrary cartesian coordinate systems in entry and exit spaces of the optical system under discussion. Let P and P' be any two points in the entry and exit spaces, respectively, and let them be in the x, y planes of their respective coordinate system.* Then the eikonal V is defined as the optical path length from P to P' along the physically possible ray joining them. In general, one and only one ray passes through P and P', but if there is more than one then V is multivalued. Thus with the above restriction V is a function of x, y, x', and y'. Let the direction cosines of the ray in the two spaces be (L, M, N) and (L', M', N'); then the fundamental property of the eikonal can be stated as follows:

$$\begin{aligned} \partial V/\partial x &= -nL, \qquad & \partial V/\partial y &= -nM \\ \partial V/\partial x' &= n'L', \qquad & \partial V/\partial y' &= n'M' \end{aligned} \tag{A.4}$$

This property is proved in many texts on geometrical optics (e.g., Welford, 1986; Born and Wolf, 1975). To prove our theorem, we differentiate Eqs. (A.4) again and we obtain, using the notations of Eq. (A.3) and using subscripts for partial derivatives,

$$\begin{aligned} dp &= -V_{xx}\,dx - V_{xy}\,dy - V_{xx'}\,dx' - V_{xy'}\,dy' \\ dq &= -V_{yx}\,dx - V_{yy}\,dy - V_{yx'}\,dx' - V_{yy'}\,dy' \\ dp' &= V_{x'x}\,dx + V_{x'y}\,dy + V_{x'x'}\,dx' + V_{x'y'}\,dy' \\ dq' &= V_{y'x}\,dx + V_{y'y}\,dy + V_{y'x'}\,dx' + V_{y'y'}\,dy' \end{aligned} \tag{A.5}$$

* This is in no way essential to the definition of the eikonal; it merely simplifies the present calculation.

We next rearrange these terms and put the equations in matrix form:

$$\begin{pmatrix} V_{xx'} & V_{xy'} & 0 & 0 \\ V_{yx'} & V_{yy'} & 0 & 0 \\ V_{x'x'} & V_{x'y'} & -1 & 0 \\ V_{y'x'} & V_{y'y'} & 0 & -1 \end{pmatrix} \begin{pmatrix} dx' \\ dy' \\ dp' \\ dq' \end{pmatrix} = \begin{pmatrix} -V_{xx} & -V_{xy} & -1 \\ -V_{yx} & -V_{yy} & 0 \\ -V_{x'x} & -V_{x'y} & 0 \\ -V_{y'x} & -V_{y'y} & 0 \end{pmatrix} \begin{pmatrix} 0\,dx \\ -1\,dy \\ 0\,dp \\ 0\,dq \end{pmatrix} \tag{A.6}$$

If we denote the two matrices by B and A and the column vectors by M and M', this equation takes the form

$$BM' = AM \tag{A.7}$$

and, multiplying through by the inverse of B,

$$M' = B^{-1}AM \tag{A.8}$$

This matrix equation can be expanded and we get together with three similar equations

$$dx' = \frac{\partial x'}{\partial x}dx + \frac{\partial x'}{\partial y}dy + \frac{\partial x'}{\partial p}dp + \frac{\partial x'}{\partial q}dq \tag{A.9}$$

It can be seen that the determinant of the matrix $B^{-1}A$ is the Jacobian

$$\det(B^{-1}A) = \frac{\partial(x', y', p', q')}{\partial(x, y, p, q)} \tag{A.10}$$

which transforms the differential four-volume $dx\,dy\,dp\,dq$, i.e., we have

$$dx'\,dy'\,dp'\,dq' = \frac{\partial(x', y', p', q')}{\partial(x, y, p, q)}\,dx\,dy\,dp\,dq \tag{A.11}$$

Our result will be proven if we can show that the Jacobian has the value unity. But the determinant of matrix B has the value

$$V_{xx'}V_{yy'} - V_{xy'}V_{yx'} \tag{A.12}$$

and that of matrix A has the same value, since $V_{xy'} = V_{y'x}$, etc. Also, the determinant of the product of two square matrices is the product of their determinants, so that

$$\det(B^{-1}) = (V_{xx'}V_{yy'} - V_{xy'}V_{yx'})^{-1} \tag{A.13}$$

Thus $\det(B^{-1}A) = 1$ and Eq. (A.13) yields our theorem, Eq. (A.2).

A.3 The Mechanical Analogies and Liouville's Theorem

In this section we shall indicate the analogies used to identify our theorem of the invariance of U, the étendue, with Liouville's theorem in statistical mechanics.

Fermat's principle, on which all of geometrical optics can be based, can be stated in the form

$$\delta \int_{P_1}^{P_2} n(x, y, z)\, ds = 0 \tag{A.14}$$

where ds is an element of the ray path from P_1 to P_2. This can be written in the form

$$\delta \int_{P_1}^{P_2} \mathscr{L}(x, y, z, \dot{x}, \dot{y})\, dz = 0 \tag{A.15}$$

where

$$\mathscr{L}(x, y, z, \dot{x}, \dot{y}) = n(x, y, z)(1+\dot{x}^2+\dot{y}^2)^{\frac{1}{2}} \tag{A.16}$$

and the dots denote differentiation with respect to z. Also we define

$$p = \frac{n\dot{x}}{(1+\dot{x}^2+\dot{y}^2)^{\frac{1}{2}}} \qquad q = \frac{n\dot{y}}{(1+\dot{x}^2+\dot{y}^2)^{\frac{1}{2}}} \tag{A.17}$$

The analogy is to regard $\mathscr{L}$ as the Lagrangian function of a mechanical system in which x and y are two generalized coordinates, p and q are the corresponding generalized momenta, and z corresponds to the time axis. On this basis the ordinary development of mechanics can be carried out as, e.g., by Luneburg (1964, Article 18), by solving the variational problem of Eq. (A.15). The Hamiltonian is found to have the value

$$\mathscr{H} = -(n^2-p^2-q^2)^{\frac{1}{2}} \tag{A.18}$$

i.e., it is $-nN$ where N is the z-direction cosine, and, of course, p and q as defined above are, respectively, equal to nL and nM. The phase space for this system has the four coordinates (x, y, p, q) and Liouville's theorem in statistical mechanics can be invoked immediately to state that phase space volume is conserved.

However, the meaning of Liouville's theorem in mechanics is rather different from the theorem of conservation of étendue. Liouville's theorem is essentially statistical in nature and it refers to the evolution in time of an *ensemble* of mechanical systems of identical properties but with different initial conditions. Each system is represented by a single point in phase space, and the theorem states that the *average* density of points in phase space is constant in time. An example would be the molecules of a perfect classical gas in equilibrium in a container. Each point in phase space, which in this example has $6N$ dimensions where N is the number of molecules, represents one of an ensemble of identical containers, an ensemble large enough to permit taking a statistical average of the density of representative points. Liouville's theorem states that if all the containers remain in equilibrium, the average density of points remains constant.

Another example would be focused beams of charged particles, as in a particle accelerator. Here we can regard one pulse of particles as constituting one realization of the ensemble, and therefore one point in phase space. The statistical averaging is carried out over the random positions and momenta of the particles entering the focusing system from pulse to pulse.

The theorem has been applied to many different physical systems but the essential point is that it makes a statistical statement about the average density of points in phase space, whereas the throughput or étendue theorem is deterministic in nature. Thus, although for convenience we may call the throughput theorem "Liouville's theorem," it is desirable to remember that there is a fundamental difference in meaning.

A.4 The Skew Invariant

In the mechanical analogy of Section A.3 the skew invariant h, introduced in Section 2.8, would be simply the "angular momentum" of the ray, and we could invoke the conservation of angular momentum to prove that h is an invariant along a ray in a system with axial symmetry. This is perhaps not quite satisfying in a book on geometrical optics, so we give below an independent proof.

Let **r** in Fig. A.1 be a skew ray in a symmetrical optical system, and let OP be the line perpendicular to the axis and the ray **r**. Thus $OP = S$, say, is the shortest distance between the ray and the axis, and the skew invariant h is $nS \sin \theta$. Since the optical system is symmetric about the axis, we can rotate the whole ray throughout the system by

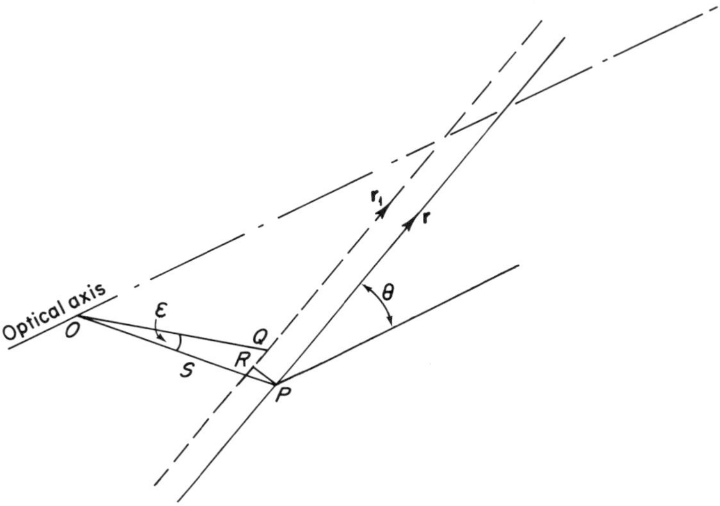

Fig. A.1 The skew invariant.

an angle ε about the axis and it will still be a possible ray in the system. Let $\mathbf{r}_1$ be this new ray, with shortest distance OQ, and define R, the foot of the perpendicular from P to the ray $\mathbf{r}_1$. We can imagine a similar diagram with primed quantities in any other space of the optical system and then we have, by axial symmetry,

$$[PP'] = [QQ'] \tag{A.19}$$

where the brackets indicate optical path lengths. Also, if ε is small, since PR is perpendicular to both rays,

$$[PP'] = [RR'] + O(\varepsilon^2) \tag{A.20}$$

Thus by subtraction

$$[RQ] = [R'Q'] + O(\varepsilon^2) \tag{A.21}$$

i.e.,

$$(nS \sin\theta)\varepsilon = (n'S' \sin\theta')\varepsilon + O(\varepsilon^2) \tag{A.22}$$

Thus by letting ε tend to zero, we have proved the invariance of h. The result is, of course, valid for systems with continuously varying refractive index.

A.5 Conventional Photometry and the Étendue

In discussing the action of concentrators we use the notion of a beam of radiation in which a certain cross section $dx\,dy$ is uniformly filled with rays covering uniformly the direction cosine solid angle $dL\,dM$. Note the useful relation $dL\,dM = N\,d\Omega$. Then an ideal concentrator takes all the rays from a source of finite area within a certain range of direction cosines and delivers them at the exit aperture to emerge over a solid angle 2π.

These ideas relate easily to particle beams but it may not be quite clear how they relate to classical photometric ideas. In classical photometry we have a particular kind of ideal source called *Lambertian* for which the flux or power radiated per unit solid angle per unit *projected* area of the source is constant over all directions. Many physical sources approximate closely to the Lambertian condition and an ideal blackbody radiator must be Lambertian since the radiation in a blackbody cavity is isotropic. It is easy to show that if a source radiates the same flux per unit area of the source and per unit element of direction cosine space, then it is Lambertian. Thus we see that an ideal concentrator will take radiation from a Lambertian source over a certain solid angle and deliver it so as to make the exit aperture appear to be a complete Lambertian radiator over solid angle 2π.

APPENDIX B

The Impossibility of Designing a "Perfect" Imaging System; the Corresponding Nonimaging Problem

The problem of the "perfect" optical system has recurred in the literature since the time of James Clerk Maxwell. Further, it has given rise to some confusion because the adjective "perfect" has been used to mean different things. We have to understand these distinctions in order to see in what sense it is impossible to design an ideal concentrator as an imaging system.

First, we classify aberrations into (1) point-imaging aberrations and (2) aberrations of image shape. If the rays from one object point do not all meet at a single image point, we have point-imaging aberrations. These can be subclassified but this is for the moment irrelevant. If there are no point-imaging aberrations for all object points over a certain area of the object plane (or a certain range of angles if the object is at infinity), it may happen that the image points fall on a curved surface rather than a plane. Then we have the aberration of image shape known as *field curvature*. If there is no field curvature and there are no point-imaging aberrations, it may happen that the image shape is not geometrically similar to that of the object and then we have *distortion*. Of course, in general all types of aberration are present and the description of a given system in these terms may present some difficulties. For example, if there are point-imaging aberrations that vary across the object, how do we define the field curvature, or if there is field curvature, how do we specify the distortion? Fortunately, we do not need to go deeply into these questions for our present purposes and the crude descriptions given above will serve us.

Maxwell (1858) considered systems that form images with neither point-imaging aberrations nor aberrations of image shape. It is easy to show that if a system forms perfect images in this sense for two longitudinally separated object planes, it will form perfect images of all object planes. Maxwell showed that it is in general impossible for an optical system to form perfect images of two different planes. The exceptions are trivial, namely combinations of plane mirrors and combinations of plane parallel plates used to image objects at infinity.

Clearly, Maxwell's definition of perfect imagery is too narrow for our purposes. We require only one object surface, usually at infinity, to be imaged without aberrations. Even then we may admit some aberrations of image shape, but this will depend on the details of what the system is to be used for.

In conventional lens design it seems to be well established by the negative results of the efforts of many skilled and experienced designers that no manufacturable system can be made to image a finite region of one real plane on another plane with no aberrations, even if we assume that chromatic aberrations do not matter. On the other hand, there are tolerances for geometrical aberrations based on unavoidable diffraction effects in lens systems. It is certainly possible to design such a system within these tolerances, and even to allow a reasonable wavelength range, i.e., apply the tolerances to chromatic aberrations. Thus in this sense perfect imaging systems are possible and are actually made and used. For example, a $10\times$ microscope objective of numerical aperture 0.25 would have substantially perfect correction over an angular field of $\pm 1^\circ$ from the axis (or $\pm 3^\circ$ if field curvature is allowed) and over a wavelength range of about 50 nm in the middle of the visible spectrum. Such optical systems are rather small and if we scale them up in size the residual aberrations scale proportionally. However, the physical aberration tolerances mentioned above are in terms of the wavelength of light, and thus the residual aberrations grow above the tolerance levels as the scale of the system is increased. For this reason alone it would not be possible to design an image-forming system perfect to within the diffraction tolerance limits on a scale suitable for solar energy collection, apart from the obvious drawbacks of cost and complexity.

As suggested in Section 3.3, if we admit extreme and practically unattainable material properties, it becomes easier to design diffraction-limited image-forming systems, but again this is of no practical use. While there is no formal proof of the impossibility of an exactly aberration-free design for imaging a plane on a plane, we can give a heuristic argument to suggest why this is so for a finite number of lens or mirror components. Let us neglect chromatic effects as depending purely on material properties. In the first place, it is easy to form a perfect image of the axial object point alone with a single thin lens, as in Fig. B.1. One of the two lens surfaces is given a nonspherical shape chosen to bring all the *axial* rays to a single point focus, a well-known technique.* If now we trace rays in the plane of the diagram from an object point O_1 away from the axis, these rays will not meet at a point. We could change the lens by adding the same increment of curvature to both surfaces (this is called "bending" the lens by optical designers) and changing the aspherizing to keep the axial pencil of rays corrected. This might improve the off-axis correction, but there is no reason to suppose that we could simultaneously get perfect correction for both O_1 and O since at any given bending the axial aspherizing uses up all available degrees of freedom.

If we add another aspheric lens to the system with a finite air space, as in Fig. B.2 (or, alternatively, we could use a single thick lens with both surfaces aspheric), this appears to give more freedom. If we attempt to write down the conditions that meridian rays from both O and O_1 will form perfect images, we obtain simultaneous differential equations for the two asphérics. Unfortunately, in general, these equations are overconstrained. Consider a region P_1 on one side of the second aspheric. The surface slope here must fulfill certain requirements to refract the rays from O and O_1 in certain directions, but the region P_2 on the other side of the axis, which necessarily has the same slope by symmetry, has to fulfill different requirements for the ray from O_1 that passes through it. Thus, in general, even the very simple requirement for an axially symmetric lens system to form perfect images of points at only one distance from the axis and by meridian rays only, i.e., rays in a plane containing the axis, cannot be fulfilled.

* These curves are called *cartesian ovals*, after Rene Descartes, who first described them; see, e.g., Luneburg (1964), Section 23.5.

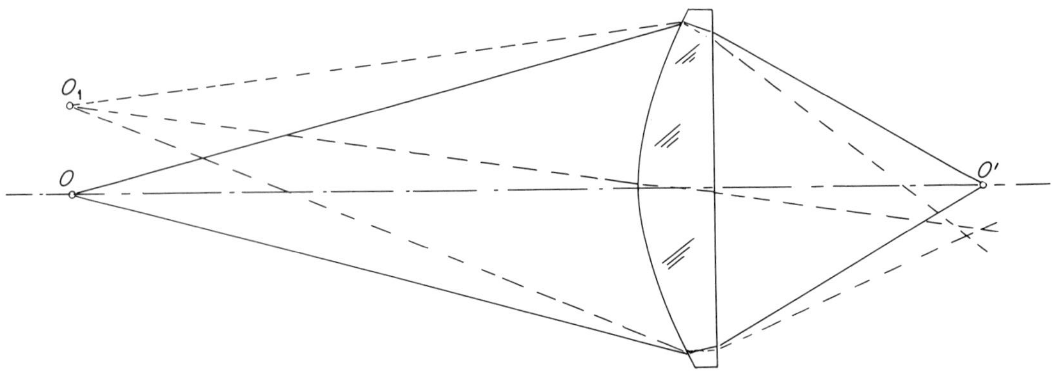

Fig. B.1 A singlet lens with the front surface aspherized to remove spherical aberration for the axial conjugates shown.

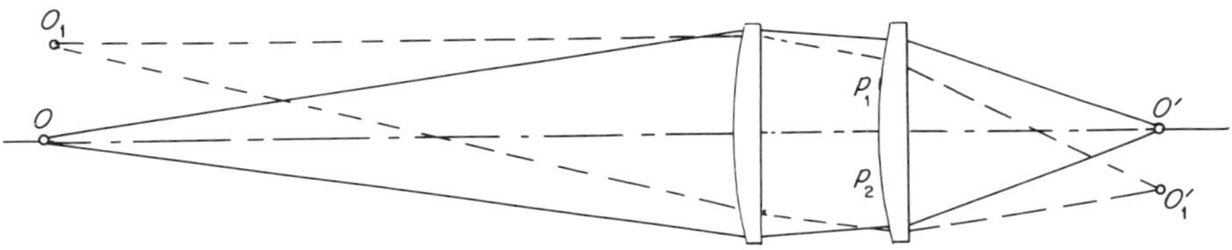

Fig. B.2 Two separated lenses, both aspherized. In general, such a combination, if axially symmetric, cannot produce aberration-free images of both O and O_1.

We could circumvent this difficulty by putting the second lens so far from the first that all rays from O_1 meet it on the same side of the axis. This is, in effect, how the reflecting concentrators achieve their results, since the extreme angle rays are operated on only by reflections on one side of the axis. However, as far as imaging systems are concerned, we are not much helped because we still have to focus the skew or nonmeridian rays from O_1, and that involves many more conditions. If this could be done, we should then have to consider other object points between O_2 and the axis! We should be led to more and more components and eventually to systems with continuously varying refractive index. Luneburg (1964) showed how this could be done, but it seems that in order to cover a continuous distribution of field points it is necessary to have a system with spherical symmetry. Clearly, perfect image-forming systems from a plane real object to a plane real image seem to be impossible.

As to nonimaging systems, we have seen already that the CPC in 3D is not ideal, but the flow-line concentrator is; however, the FLC has the disadvantage of a virtual entry aperture and it is necessary to add lenses to produce a real entry aperture. An interesting variant on the CPC theme is a crossed or rectangular CPC, as in Fig. B.3; this consists of sections of 2D CPCs at right angles and it might be thought that this would form an ideal 3D system, albeit not axisymmetric. In fact, it was found by ray tracing that this is not so; some rays are turned back. Nevertheless, it is a useful system for some purposes, notably in reverse for collecting light from rectangular sources (see Section 11.4).

We may end by quoting a curious result due to Garwin (1952), who showed that, in our terminology, a 3D concentrator can be designed to transform an area of any shape into any other while conserving

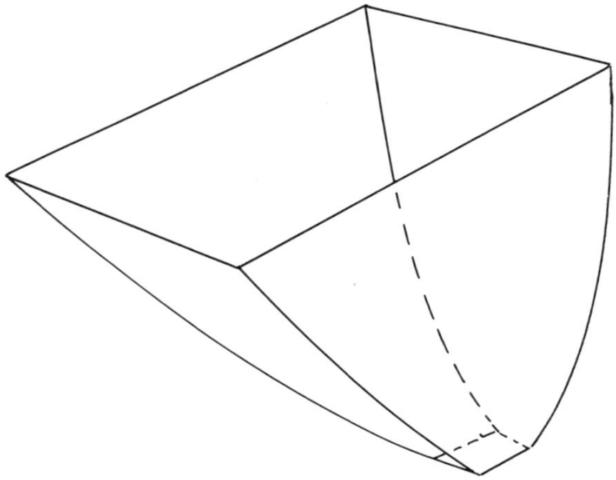

Fig. B.3 A 3D concentrator constructed from sections of 2D concentrators.

étendue, but that it is necessary to start the concentrator as a cylindrical surface and to change its shape adiabatically. In effect, this means that the concentrator would have to be infinitely long to achieve any concentration greater than unity!

APPENDIX C

The Luneburg Lens

The Luneburg lens, discussed in Chapter 3 of this book, is not, of course, of any use as a practical concentrator. It is, however, the simplest example of an ideal concentrator with maximum theoretical concentration for collecting angles up to $\pi/2$, and therefore we develop its theory and general properties in this Appendix. The main reference is to Luneburg's own work (Luneburg, 1964). Some of the general geometrical optics background can also be found in Born and Wolf (1975), and extensions are given by Morgan (1958) and Cornbleet (1976).

Our starting point is the differential equation of the light rays. Let s be distance along a ray measured from some fixed origin, let $\mathbf{r}$ be the position vector of a point on a ray, and let $n(\mathbf{r})$ be the refractive index, varying continuously as a function of $\mathbf{r}$. Then it can easily be shown [e.g., Born and Wolf (1975, Section 3.2.1)] that the rays are given by solutions of

$$\frac{d}{ds}\left(n(\mathbf{r})\frac{d\mathbf{r}}{ds}\right) = \text{grad } n(\mathbf{r}) \tag{C.1}$$

Let the refractive index distribution have spherical symmetry. Then we can write $n(\mathbf{r}) \equiv n(r)$ where r is a radial coordinate from the origin. Also, on account of the spherical symmetry, any light ray lies wholly in a plane through the origin and we can describe its path by the polar coordinates (r, θ).

In order to do this, we note that $d\mathbf{r}/ds$ is a unit vector along the tangent to the ray, say $\mathbf{s}$. Now we have

$$\frac{d}{ds}(\mathbf{r}\times n\mathbf{s})=\mathbf{s}\times n\mathbf{s}+\mathbf{r}\times\frac{d(n\mathbf{s})}{ds}=\mathbf{r}\times\operatorname{grad} n$$

from Eq. (C.1). But by spherical symmetry grad n is parallel to $\mathbf{r}$ and so we have

$$\frac{d}{ds}(\mathbf{r}\times n\mathbf{s})=0$$

and a first integral of Eq. (C.1) is

$$\mathbf{r}\times n\mathbf{s}=\text{const.}$$

Since the rays are plane curves as noted above, this gives, say,

$$nr\sin\varphi=\text{const.}=h \tag{C.2}$$

where φ is as in Fig. C.1. Since

$$\tan\varphi=r\,d\theta/dr$$

Eq. (C.2) yields

$$d\theta/dr=h/r(n^2r^2-h^2)^{1/2} \tag{C.3}$$

and thus if $n(r)$ is specified, the ray paths are obtained by quadrature:

$$\theta_1-\theta_0=h\int_{r_0}^{r_1}\frac{dr}{r(n^2r^2-h^2)^{1/2}} \tag{C.4}$$

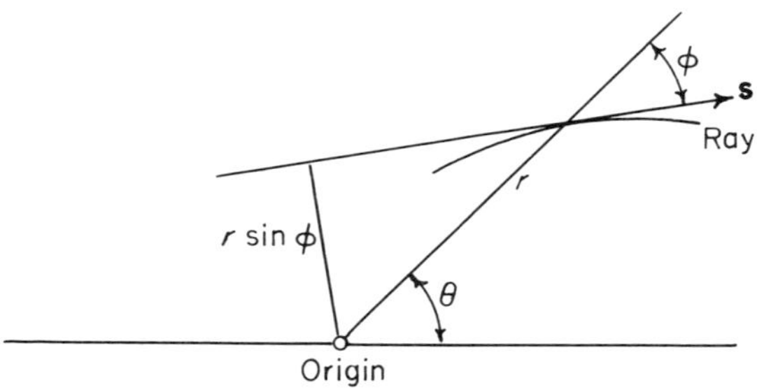

Fig. C.1 Notation for ray paths in a medium with spherical symmetry.

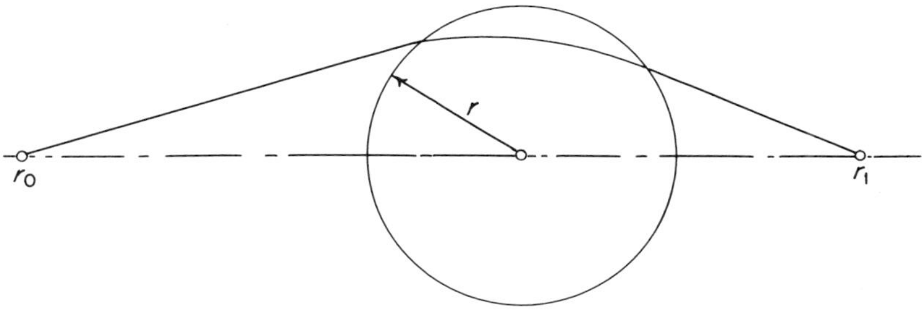

Fig. C.2 Luneburg's general problem of an aberrationless unit sphere.

If we wish to find an index distribution $n(r)$ that will act as an aberration-free medium, we have to regard Eq. (C.4) as an integral equation for the function $n(r)$. Luneburg did this and considered the use of an index distribution inside the unit sphere and with index unity at the surface of the sphere. He formulated the conditions for focusing from any point r_0 to another r_1 as in Fig. C.2 and he obtained an explicit solution for the case of present interest, $r_0 = \infty$, $r_1 = 1$. Luneburg's derivation is very complex and closely argued. Rather than reproduce his work we shall simply verify the solution he gave.

Let the lens have radius a and let the index vary in such a way that r has only one minimum value for the ray path in the lens, say, r^*, as in Fig. C.3. Then from Eq. (C.2) we have, using values at the

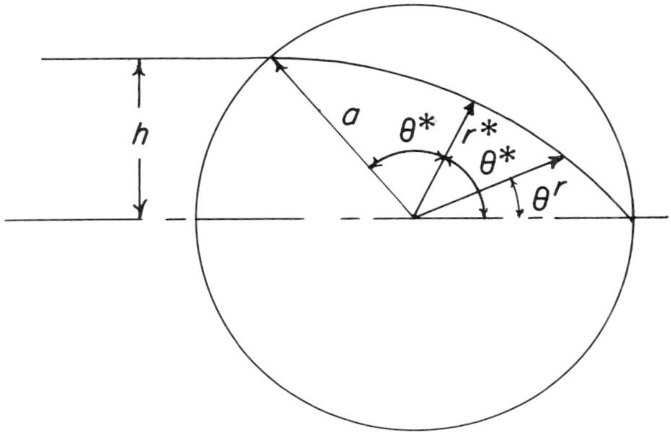

Fig. C.3 Luneburg's problem for $r_0 = \infty$, $r_1 = 1$.

beginning of the trajectory in the lens,

$$nr \sin \varphi \equiv a \cdot h/a = h$$

and thus

$$n(r^*)r^* = h \tag{C.5}$$

Since the two halves of the trajectory on either side of the radius r^* must be mirror images, we have

$$\theta^* = \pi/2 - \tfrac{1}{2} \arcsin(h/a) \tag{C.6}$$

and thus Eq. (C.4) takes the form

$$\pi/2 - \tfrac{1}{2} \arcsin(h/a) = h \int_{r^*}^{a} \frac{dr}{r(n^2r^2 - h^2)^{1/2}} \tag{C.7}$$

to be satisfied by a function $n(r)$ for any $h < a$.

Luneburg gave the distribution

$$n(r) = (2 - r^2/a^2)^{\frac{1}{2}} \tag{C.8}$$

as the solution of this integral equation. If we substitute for $n(r)$, we have to evaluate

$$h \int_{r^*}^{a} \frac{dr}{r(2r^2 - r^4/a^2 - h^2)^{\frac{1}{2}}} \tag{C.9}$$

to verify the solution. From Eq. (C.5) we find

$$r^{*2} = a^2[1 - (1 - h^2/a^2)^{\frac{1}{2}}] \tag{C.10}$$

and if we make the change of variable

$$r^2 = a^2[1 - (1 - h^2/a^2)^{1/2} \sin \beta] \tag{C.11}$$

in the integral, it becomes

$$h \int_{0}^{\pi} \frac{d\beta}{2a[1 - (1 - h^2/a^2)^{\frac{1}{2}} \sin \beta]} \tag{C.12}$$

The indefinite integral has the value

$$\tan^{-1} \left\{ \frac{\tan \frac{1}{2}\beta - (1 - h^2/a^2)^{\frac{1}{2}}}{h/a} \right\} \tag{C.13}$$

and on substituting the limits of integration and recalling that $h/a = \sin 2\theta^*$ we find that the value is θ^*, as required by Eq. (C.7). We have

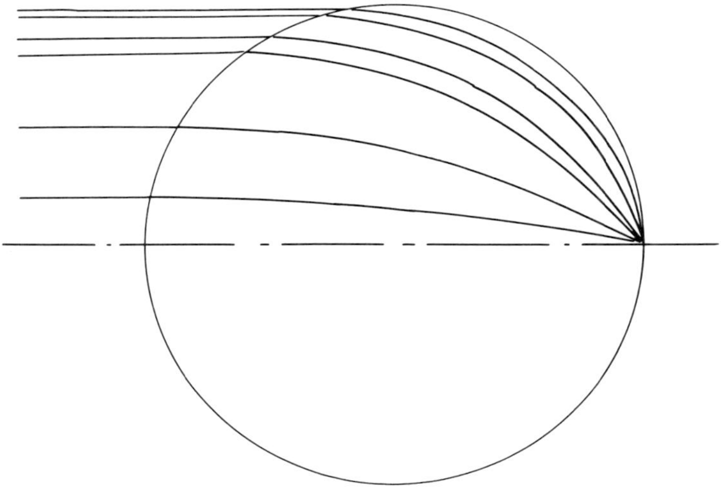

Fig. C.4 Rays in the Luneburg lens.

also obtained the actual ray paths, since these are specified by θ as a function of r for given h by

$$\theta_1 - \theta_0 = \tan^{-1}\left\{\frac{\tan\frac{1}{2}\beta - (1 - h^2/a^2)^{\frac{1}{2}}}{h/a}\right\}\Bigg|_{r_0}^{r_1} \tag{C.14}$$

with $\sin\beta = [1 - (r^2/a^2)](1 - h^2/a^2)^{-\frac{1}{2}}$. Figure C.4 shows several rays plotted to scale.

In order to use the Luneburg lens as an ideal concentrator, we choose an input angle θ_i less than $\pi/2$ and use the opposite spherical cap as the absorbing surface, as in Fig. C.5. A ray incident at height h from the center emerges at an angle to the surface arcsin (h/a), so that at any point on the exit surface rays emerge in all directions up to $\pi/2$ from the normal and the sines of the angles are distributed uniformly. Thus the exit surface is filled with rays at all possible angles and this must be in some sense a system with maximum theoretical concentration ratio. We can make this more explicit by calculating the entering étendue. This is, from Fig. C.5, the area $\pi a^2 \sec\theta$ integrated over the direction cosines of the incoming beam, i.e., it is

$$\int_0^{\theta_1}\int_0^{2\pi} \pi a^2 \sec\theta \, d(\sin\theta\cos\varphi)\, d(\sin\theta\sin\varphi)$$

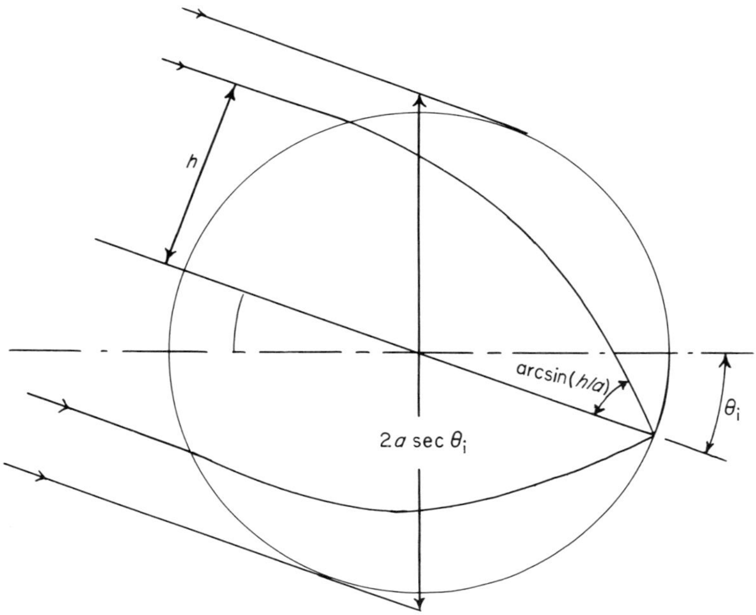

Fig. C.5 The Luneburg lens as a concentrator of maximum theoretical concentration ratio for any entry angle θ_i less than $\pi/4$.

where φ is an azimuthal angle. The value of this is $2\pi^2 a^2(1-\cos\theta_i)$. Also, the integral over the direction cosines alone is

$$\int_0^{\theta_i}\int_0^{2\pi} d(\sin\theta\cos\varphi)\,d(\sin\theta\sin\varphi)$$

and this is $\pi\sin^2\theta_i$. Thus the effective entry area must be $\pi a^2/\cos^2\frac{1}{2}\theta_i$.

The area of the spherical cap that forms the exit surface is $4\pi a^2\sin^2\frac{1}{2}\theta_i$ so that the concentration ratio is $1/\sin^2\theta_i$, the theoretical value for a collecting angle θ_i. Since all entering rays emerge, we have a 3D concentrator with maximum theoretical concentration ratio.

We had to proceed in the above slightly oblique way because the collecting aperture of the lens shifts with angle and because the exit surface is not plane. Thus the correspondence with systems with well-defined and plane entry and exit apertures is not perfect.

APPENDIX D

The Geometry of the Basic Compound Parabolic Concentrator

It is probably simplest to obtain the basic properties of the CPC from the equation of the parabola in polar coordinates. Figure D.1 shows the coordinate system. The focal length f is the distance AF from the vertex to the focus. The equation of the parabola is, then,

$$r = 2f/(1 - \cos\varphi) = f/(\sin^2 \tfrac{1}{2}\varphi) \tag{D.1}$$

This result is given in elementary texts on coordinate geometry and it may be verified by transforming to polars the more familiar cartesian form, $z = y^2/4f$ with axes as indicated in the figure.

We apply this to the design of the CPC as in Fig. D.2. We first draw the entrance and exit apertures PP' and QQ' with the desired ratio in aperture between them, and we choose the distance between them so that an extreme ray PQ' (or $P'Q$) makes the maximum collecting angle θ_i with the concentrator axis. Then according to Section 4.3 the profile of the CPC between P' and Q' is a parabola with axis parallel to PQ' and with focus at Q, and this parabola can be expressed in terms of the polar coordinates (r, φ) as in the diagram.

For the exit aperture we have, using Eq. (D.1),

$$QQ' = 2f/[1 - \cos(\pi/2 + \theta_i)]$$

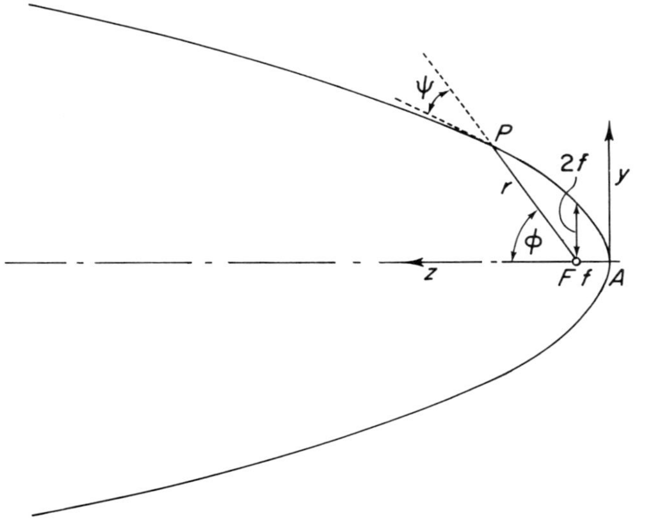

Fig. D.1 The parabola in polar coordinates with origin at the focus. $r = 2f(1-\cos\phi) = f\sin^2\frac{1}{2}\phi$.

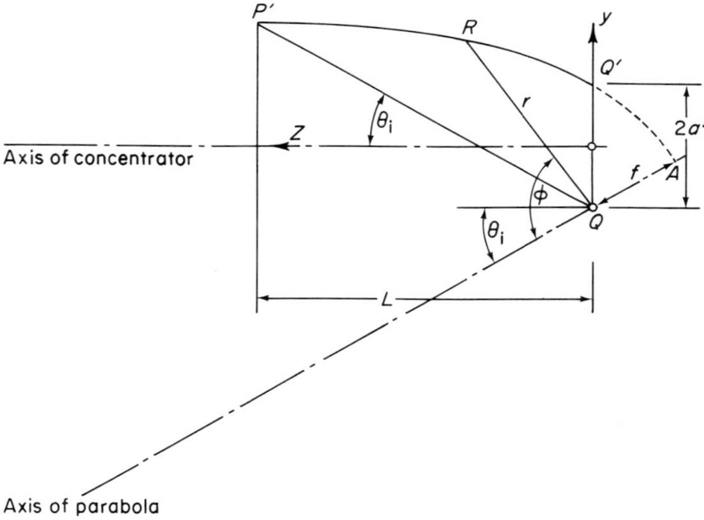

Fig. D.2 Design of the CPC.

so that, if $QQ' = 2a'$, in our usual notation

$$f = a'(1 + \sin\theta_i) \tag{D.2}$$

Next we find

$$QP' = 2f/(1 - \cos 2\theta_i)$$

or

$$QP' = a'(1 + \sin\theta_i)/\sin^2\theta_i \tag{D.3}$$

Thus

$$a + a' = QP' \sin\theta_i = a'(1 + \sin\theta_i)/\sin\theta_i$$

so that $a = a'/\sin\theta_i$, as required. Finally, for the length of the concentrator

$$L = QP' \cos\theta_i = a'(1 + \sin\theta_i) \cot\theta_i/\sin\theta_i = (a + a') \cot\theta_i \tag{D.4}$$

This, of course, is again as required by the way we set up the design of the CPC.

It is obvious from the geometry of the initial requirements for the CPC that the profile at P and P' should have its tangent parallel to the axis. This also can be verified from the above formulation. For the tangent of the angle between the curve and the radius vector is, in polars,

$$\tan\psi = r\,d\varphi/dr = 2f/(r \sin\varphi)$$

for the parabola, and if we put $r = QP'$ and $\varphi = 2\theta_i$, we find

$$\tan\psi = \tan\theta_i$$

as expected.

The parametric representation of the CPC profile given in Section 4.3 is easily obtained from Fig. D.2 if we take an origin for cartesian coordinates at the center of the exit aperture and z axis along the concentrator axis. We have from the figure

$$y = r \sin(\varphi - \theta_i) - a' = \frac{2f \sin(\varphi - \theta_i)}{1 - \cos\varphi} - a'$$

$$= \frac{2a'(1 + \sin\theta_i) \sin(\varphi - \theta_i)}{1 - \cos\phi} - a'$$

$$z = r \cos(\varphi - \theta_i) = \frac{2a'(1 + \sin\theta_i) \cos(\varphi - \theta_i)}{1 - \cos\varphi}$$

APPENDIX E

The θ_i/θ_o Concentrator

Figure E.1 shows the concentrator with appropriate notation. From the triangle $QQ'S$ we have

$$QS = 2a' \cos \tfrac{1}{2}(\theta_o - \theta_i)/\sin \tfrac{1}{2}(\theta_o + \theta_i) \tag{E.1}$$

Also, since the angle $R\hat{Q}S$ is $\pi - \theta_o - \theta_i$, the triangle RQS is isosceles and so $QR = QS$.

From the polar equation of the parabola,

$$QR = 2f/[1 - \cos(\theta_o + \theta_i)] \tag{E.2}$$

so from Eq. (E.1) we find for the focal length of the parabola

$$f = 2a' \cos \tfrac{1}{2}(\theta_o - \theta_i) \sin \tfrac{1}{2}(\theta_o + \theta_i) = a'(\sin \theta_i + \sin \theta_o) \tag{E.3}$$

Again using the polar equation of the parabola,

$$QP' = \frac{2f}{1 - \cos 2\theta_i} = a'(\sin \theta_i + \sin \theta_o)/\sin^2 \theta_i$$

and

$$a + a' = QP' \sin \theta_i = \frac{a'(\sin \theta_i + \sin \theta_o)}{\sin \theta_i} \tag{E.4}$$

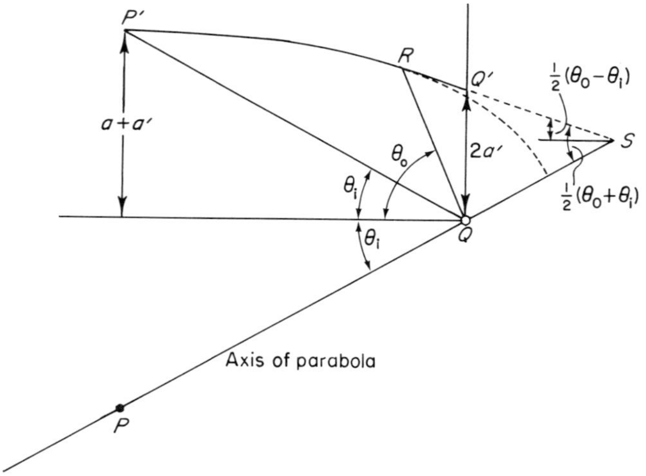

Fig. E.1 Design of the θ_i/θ_o concentrator.

From Eq. (E.4) we find immediately

$$a/a' = \sin\theta_o/\sin\theta_i \tag{E.5}$$

so that the θ_i/θ_o concentrator is actually ideal in two dimensions.

The overall length of the concentrator is, from the figure,

$$L = (a + a')\cot\theta_i \tag{E.6}$$

just as for the basic CPC [Eq. (D.4)].

APPENDIX F

The Concentrator Design for Skew Rays

F.1 The Differential Equation

The profile of the modified concentrator described in Section 5.6 can be described by means of the coordinate system in Fig. F.1. The origin is taken at the center of the exit aperture and this latter has radius a' as usual. We trace a ray in reverse from some point Q on the exit aperture to the point $R(0, y, z)$ on the desired profile and then reflect it. This ray must be constrained to leave Q with the chosen value h of the skew invariant and after reflection it must have the inclination θ_i to the axis. These constraints will determine the profile uniquely.

If Q is $(x_0, y_0, 0)$, the direction cosines of the ray QR are

$$(\alpha, \beta, \gamma) = (-x_0, y - y_0, z)/(x_0^2 + (y - y_0)^2 + z^2)^{\frac{1}{2}} \tag{F.1}$$

with the condition

$$x_0 y/(x_0^2 + (y - y_0)^2 + z^2)^{\frac{1}{2}} = h \tag{F.2}$$

Let the inclination of the normal at R be ψ, as indicated in the figure. Then the law of reflection [Eq. (2.1)] gives for the direction cosines of the reflected ray RP

$$(\alpha', \beta', \gamma') = (\alpha, -\beta \cos 2\psi + \gamma \sin 2\psi, \beta \sin 2\psi + \gamma \cos 2\psi) \tag{F.3}$$

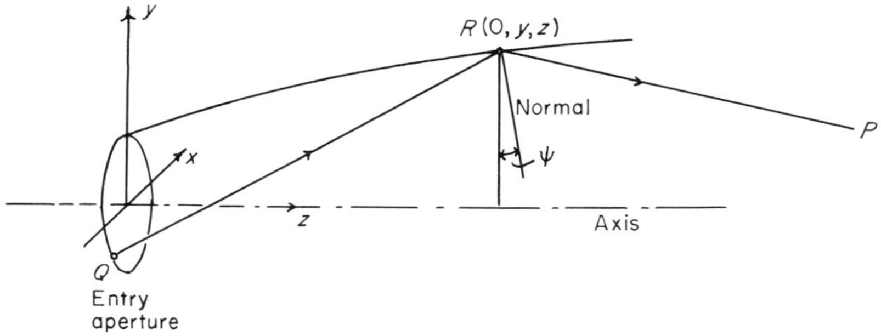

Fig. F.1 Constructing the differential equation for the concentrator designed for skew rays.

Our second constraint requires that $\gamma' = \cos\theta_i$, i.e.,

$$\gamma\cos 2\psi + \beta\sin 2\psi = \cos\theta_i \tag{F.4}$$

Now $\tan\psi = dy/dz$, the slope of the required profile, so Eq. (F.4) takes the form

$$\gamma[1-(dy/dz)^2] + 2\beta(dy/dz) = \cos\theta_i[1+(dy/dz)^2] \tag{F.5}$$

This can be solved for dy/dz since β and γ are functions of y and z through Eqs. (F.1) and (F.2), and the result is a first-order nonlinear differential equation for the profile.

The actual solution must, of couse, be done numerically.

F.2 The Ratio of Input to Output Areas for the Concentrator

The process explained in Section F.1 determines a unique profile for given h and θ_i but nothing was put into the solution to ensure that the ratio of entry to exit areas has the correct value, $(1/\sin\theta_i)^2$. In other words, the length of the concentrator would be determined by where a ray starting in the plane of the exit aperture with skew invariant h meets the surface, and there is no reason why the diameter so found should equal $2a\sin\theta_i$. In fact, it turns out that this is always so, so that this concentrator could, if all rays were transmitted, have maximum theoretical concentration. We prove this result in this section.

We use in the proof a method of treating skew rays in an axisymmetric system due to Luneburg (1964). In this method a skew ray in a part of the system with uniform index of refraction is treated as a meridian ray in the same part of the system but with a radially varying index. We first derive this formalism. If (ρ, θ, z) are cylindrical coordinates, then the optical path length along a ray between planes z_0 and z_1 is, for any distribution of index $n(\rho, z)$,

$$V = \int_{z_0}^{z_1} n(\rho, z)(1+\dot{\rho}^2+\rho^2\dot{\theta}^2)^{\frac{1}{2}}\, dz \tag{F.6}$$

where the dots indicate differentiation with respect to z. In terms of these quantities it is easily seen that the skew invariant is

$$h = n\rho^2\dot{\theta}/(1+\dot{\rho}^2+\rho^2\dot{\theta}^2)^{\frac{1}{2}} \tag{F.7}$$

From Eq. (F.6)

$$\begin{aligned} V &= \int_{z_0}^{z_1} \frac{n(1+\dot{\rho}^2+\rho^2\dot{\theta}^2)}{(1+\dot{\rho}^2+\rho^2\dot{\theta}^2)^{\frac{1}{2}}}\, dz \\ &= \int_{z_0}^{z_1} \frac{h(1+\dot{\rho}^2+\rho^2\dot{\theta}^2)}{\rho^2\dot{\theta}}\, dz \\ &= \int_{z_0}^{z_1} \frac{h(1+\dot{\rho}^2)}{\rho^2\dot{\theta}}\, dz + h(\theta_1-\theta_0) \end{aligned} \tag{F.8}$$

Now, we can solve Eq. (F.7) for $\dot{\theta}$, giving

$$\dot{\theta} = \frac{h(1+\dot{\rho}^2)^{1/2}}{\rho(n^2\rho^2-h^2)^{1/2}}$$

and if this is substituted in the integral in Eq. (F.8), we obtain

$$V = \int_{z_0}^{z_1} \left(n^2 - \frac{h^2}{\rho^2}\right)^{1/2} (1+\dot{\rho}^2)^{\frac{1}{2}}\, dz + h(\theta_1-\theta_0) \tag{F.9}$$

This result, due to Luneburg, has the significance that the optical path length along a skew ray between planes z_0 and z_1 can be computed as that along a meridian ray between $z_0 z_1$ in a medium of "refractive index"

$$m(\rho, z) = [n^2(\rho, z) - h^2/\rho^2)^{1/2} \tag{F.10}$$

together with an azimuthal contribution $h(\theta_1-\theta_0)$. Let the longitudinal and azimuthal components of the right-hand side of Eq. (F.9) be denoted by V_z and V_θ. Then the result shows that in the ρ, z plane the "rays" obey Fermat's principle for V_z with the fictitious index m given by Eq. (F.10), so that there is true image formation in this plane.

Some rays in the concentrator designed for nonzero h appear as in Fig. F.2 in the Luneburg representation. We show rays entering with the collecting angle θ_i and parallel to each other. Ray 1 strikes the entry aperture at A and is reflected to the rim of the exit aperture at B, while ray 2 grazes the entry aperture at D and is reflected at the exit aperture at E. Ray 2 has, from the design, the skew invariant h, so that it travels perpendicular to the axis and meets the exit aperture again at C. Points B and C are in fact coincident, but we have not yet shown this. We wish to calculate the optical path lengths V_{z_1} and V_{z_2} from the source point at infinity along rays 1 and 2 to points B and C on the exit rim. However, since rays 1 and 2 are parallel before they enter the concentrator, it suffices to compute the paths from a plane through D perpendicular to both rays. We have for these paths

$$\begin{aligned} V_1 &= 2a \sin \varphi_1 \sin \theta_i + V_1(AB) \\ V_2 &= V_2(DE) + 2a' \sin \varphi_2 \end{aligned} \tag{F.11}$$

and $V_1(AB) = V_2(DE)$, since these two paths are geometrically identical.

Now, in the Luneburg representation the longitudinal path lengths from infinity to the exit rim are equal, since our solution of the differential equation in Section F.1 has ensured this. Likewise, the longitudinal path lengths from A to B and from D to E are equal, so that

$$V_{z_1} = V_{z_2}$$

or

$$V_1 - h\theta_1 = V_2 - h\theta_2 \tag{F.12}$$

where θ_1 and θ_2 are the appropriate azimuthal angles for the two paths. Combining Eqs. (F.12) and (F.12), we obtain

$$V_1 - V_2 = 2a \sin \varphi_1 \sin \theta_i - 2a' \sin \varphi_2 = h(\theta_1 - \theta_2) \tag{F.13}$$

If we compute the angles θ_1 and θ_2 from infinity to B and C, respectively, and remember that AB and DE subtend the same angles, we find

$$\theta_1 - \theta_2 = 2(\varphi_1 - \varphi_2) \tag{F.14}$$

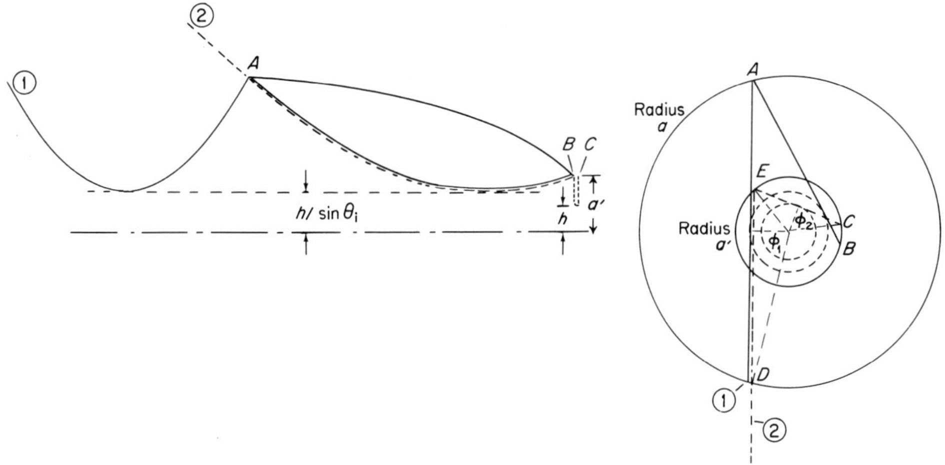

Fig. F.2 Rays in the Luneburg representation.

and also from the skew invariant we find

$$\cos \varphi_1 = h/(a \sin \theta_i), \qquad \cos \varphi_2 = h/a' \tag{F.15}$$

We use Eqs. (F.14) and (F.15) to eliminate a, a', θ_i, and θ_2 from Eq. (F.13) and obtain

$$\tan \varphi_1 - \tan \varphi_2 = \varphi_1 - \varphi_2 \tag{F.16}$$

Now $\tan \varphi - \varphi$ is monotonic for φ less than $\pi/2$ and the angles in Eq. (F.16) are less than $\pi/2$, so we conclude that $\varphi_1 = \varphi_2$ and thus, from Eq. (F.15),

$$a'/a = \sin \theta_i \tag{F.17}$$

Thus the concentrator designed for the nonzero skew invariant has the ratio of apertures required to give the maximum theoretical concentration ratio.

We can also calculate the length L of the concentrator:

$$L = [(a^2 - h^2/\sin^2 \theta_i)^{\frac{1}{2}} + (a'^2 - h^2/\sin^2 \theta_i)^{\frac{1}{2}}] \cot \theta_i \tag{F.18}$$

F.3 Proof That Extreme Rays Intersect at the Exit Aperture Rim

We stated in Section 5.6 in connection with Fig. 5.8 that rays in a chordal plane at θ_i and touching either side of the entry rim (points A and D in Fig. 5.8) will emerge at the same point C of the exit aperture rim. We can show this result very simply, as in Fig. F.3; this is the same as Fig. 5.8 but with some construction lines added.

We assume that the ray reflected from B meets the exit rim at C and that reflected from D meets it at F, and we shall show that F and C coincide. First, by construction the triangles AOB and DOF are congruent so that $\angle GOB = \angle DOF$. Thus the external angle from G to F is given by

$$\angle GOF = \angle 2GOB + \angle BOD \tag{F.19}$$

Next, the triangles AOD and BOC are similar, as may be seen from the radii of the pairs of circles involved in the construction; the ratio of sides is $\sin \theta_i$. Thus for the external angle from G to C we have

$$\begin{aligned} \angle GOC &= \angle GOB + \angle BOC = \angle GOB + \angle GOD \\ &= \angle 2GOB + \angle BOD \end{aligned} \tag{F.20}$$

Thus comparing Eqs. (F.19) and (F.20) we have proved our result.

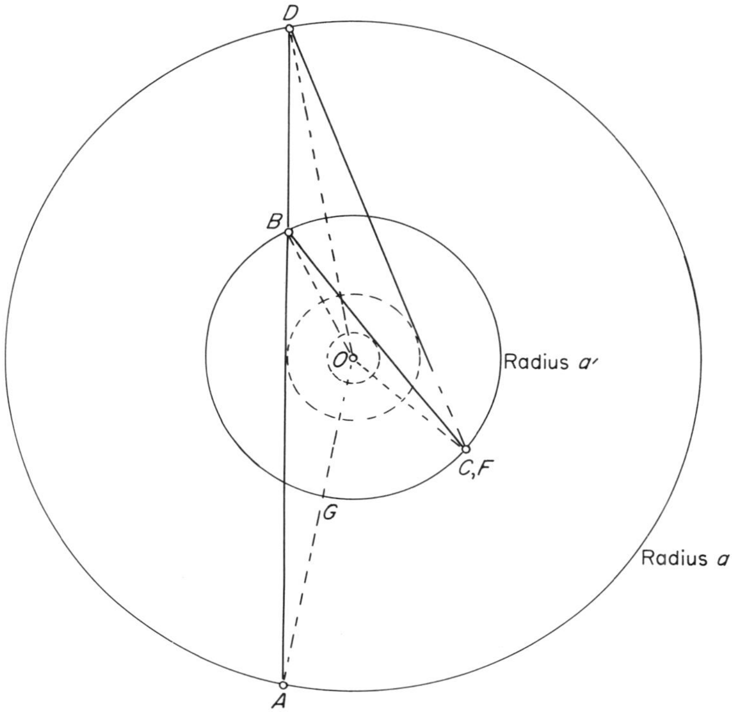

Fig. F.3 Coincidence of the points C and F at which two rays emerge.

F.4 Another Proof of the Sine Relation for Skew Rays

We may also obtain the sine relation for the concentrator designed for skew rays by formally evaluating the étendue in the p, ρ plane using the fictitious index m (bearing in mind that in these variables the concentration has ideal properties). We need to calculate

$$H = \int\int dp\, d\rho \tag{F.21}$$

where p is the optical direction cosine in the ρ direction,

$$p = m(\rho, z)\dot{\rho}/(1+\dot{\rho}^2)^{\frac{1}{2}} \tag{F.22}$$

and $m(\rho, z)$ is given by Eq. (F.10).

To find the limits on p, the following identity is useful:

$$m/(1+\dot{\rho}^2)^{\frac{1}{2}} = n \cos\theta \tag{F.23}$$

where θ is the inclination of the ray to the z axis. Then the étendue over the entrance aperture is

$$H = 2\int_{h/\sin\theta_i}^{a} (\sin^2\theta_i - h^2/\rho^2)^{\frac{1}{2}}\, d\rho \tag{F.24}$$

while the étendue over the exit aperture is

$$H' = 2\int_{h}^{a'} (1 - h^2/\rho^2)^{\frac{1}{2}}\, d\rho \tag{F.25}$$

Performing the integrations and setting $H = H'$, we obtain

$$2h[\tan\varphi_1 - \varphi_1] = 2h[\tan\varphi_2 - \varphi_2] \tag{F.26}$$

the same relation as Eq. (F.16). Therefore, the sine relation is proved.

F.5 The Frequency Distribution of *h*

Having found a solution for fixed h, we may ask for the frequency of occurrence of h in an ideal 3D system with rotational symmetry. This is readily found by writing the étendue in cylindrical coordinates (ρ, θ, z):

$$H = \int dh\, dp\, d\rho\, d\theta \tag{F.27}$$

Here p is the generalized momentum conjugate to ρ obtained from differentiating the integrand of Eq. (F.6) with respect to $\dot{\rho}$. Then h is distributed according to

$$dH/dh = 2\pi \int dp\, d\rho \tag{F.28}$$

We can just as well work in the Luneburg representation and obtain p from Eq. (F.9) (the two are of course identical). The desired integration is therefore already performed in the previous section and we may take over the result:

$$dH/dh = 4\pi h[\tan\varphi_2 - \varphi_2] \tag{F.29}$$

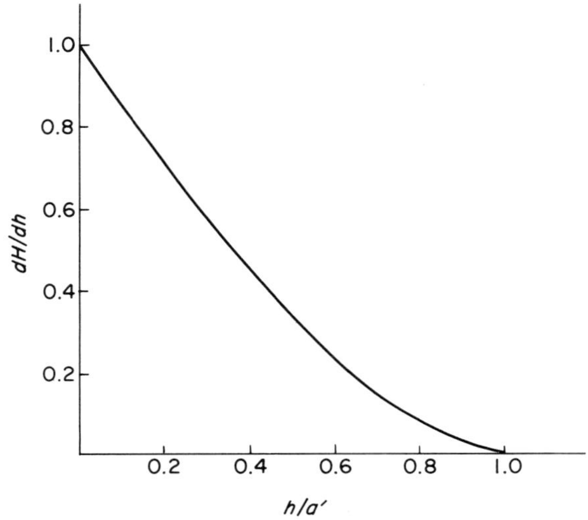

Fig. F.4 Frequency distribution in h for an ideal 3D system. The ordinate scale is chosen to have maximum value = 1.

For purposes of plotting, this expression is more conveniently written

$$dH/dh = 4\pi a'[(1-x^2)^{\frac{1}{2}} - x\cos^{-1} x] \tag{F.30}$$

where $x = h/a'$ lies in the interval $(-1, 1)$. This function is plotted in Fig. F.4. We observe that the frequency distribution strongly peaks toward $h = 0$. We may check that the distribution is properly normalized by integrating Eq. (F.30) over all h values:

$$H = \int_{-a'}^{a'} (dH/dh)dh = \pi^2 a'^2 \tag{F.31}$$

We recognize this to be the étendue of a diffusely illuminated disk of radius a'.

APPENDIX G

The Truncated Compound Parabolic Concentrator

In this appendix we give the derivations of the formulas in Section 5.7. We show in Fig. G.1 the parabolic section in polar coordinates. For these coordinates we recall (see Appendix D) that the equation of the parabola is

$$r = 2f/(1-\cos\varphi) = f/\sin^2 \tfrac{1}{2}\varphi$$
$$f = a'(1+\sin\theta_i) \qquad \text{(G.1)}$$

If s is the arc length of the parabola then

$$ds^2 = dr^2 + r^2\, d\varphi^2$$

so that

$$ds/d\varphi = [r^2 + (dr/d\varphi)^2]^{\frac{1}{2}}$$

or

$$ds/d\varphi = f/\sin^3 \tfrac{1}{2}\varphi \qquad \text{(G.2)}$$

from Eq. (G.1). We have to integrate this to find the arc length, which is proportional to the reflector area in a 2D system. We have to evaluate the indefinite integral

$$s = f\int d\varphi/\sin^3 \tfrac{1}{2}\varphi$$

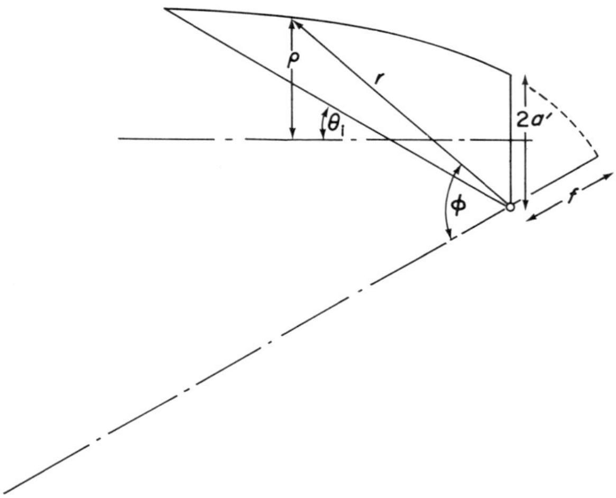

Fig. G.1 Construction for the truncated CPC.

On making the change of variable

$$\cosh u = \operatorname{cosec} \tfrac{1}{2}\varphi$$

we obtain

$$s_{\mathrm{T}} = -f\{(\cos \tfrac{1}{2}\varphi / \sin^2 \tfrac{1}{2}\varphi) + \ln \cot \tfrac{1}{4}\varphi\}\Big|_{\varphi_{\mathrm{T}}}^{\theta_{\mathrm{i}} + \pi/2} \tag{G.3}$$

The radius of the entrance aperture is simply

$$a_{\mathrm{T}} = r \sin(\varphi - \theta_{\mathrm{i}}) - a'$$

which gives Eq. (5.18) immediately. Thus Eq. (5.22) for $s_{\mathrm{T}}/a_{\mathrm{T}}$ is obtained. Likewise the length L_{T} of the truncated CPC is

$$L_{\mathrm{T}} = r \cos(\varphi - \theta_{\mathrm{i}})$$

from which Eq. (5.19) follows immediately, and thence Eq. (5.21).

For the 3D CPC the element of area of reflector is

$$dA = 2\pi\rho\, ds$$

where ρ is the radius of the CPC at the current point, as in Fig. F.1. Since ρ is $r \sin(\varphi - \theta_{\mathrm{i}}) - a'$, we have

$$dA - 2\pi \left\{ \frac{(f \sin(\varphi - \theta_{\mathrm{i}})}{\sin^2 \tfrac{1}{2}\varphi} - a' \right\} \frac{f}{\sin^3 \tfrac{1}{2}\varphi}\, d\varphi$$

This can be integrated by making the same change of variable, $\cosh u = \operatorname{cosec}\frac{1}{2}\varphi$, and the result for the ratio of area of collection surface divided by area of collecting aperture is:

$$\frac{2f}{a_T^2}\left[f\sin\theta_i\left\{\frac{3}{4}\ln\cot\frac{1}{4}\phi+\frac{3}{4}\frac{\cos\frac{1}{2}\phi}{\sin^2\frac{1}{2}\phi}+\frac{1}{2}\frac{\cos\frac{1}{2}\phi}{\sin^4\frac{1}{2}\phi}\right\}\right.$$
$$\left.-(2f\sin\theta_i-a')\left\{\ln\cot\frac{1}{4}\phi+\frac{\cos\frac{1}{2}\phi}{\sin^2\frac{1}{2}\phi}\right\}-\frac{4f\cos\theta_i}{3\sin^3\frac{1}{2}\phi}\right]_{\phi_T}^{\theta_i+\pi/2} \tag{G.4}$$

APPENDIX H

The Differential Equation for the 2D Concentrator Profile with Nonplane Absorber

It was shown in Chapter 6 that the concentrator profile that gives the theoretical maximum concentration is obtained in two sections. The first section, that shadowed from the direct rays at angles less than θ_i, the maximum input angle, is an involute of the absorber cross section. The second section is such that rays at θ_i are tangent to the absorber after one reflection at the concentrator surface.

We suppose the absorber surface to be specified by its polar coordinates r, θ, as in Fig. H.1. The current point P on the profile is at a distance t from the corresponding tangent point on the absorber. The condition for reflection is

$$\theta_i + \theta + \varphi + 2\psi = 2\pi \tag{H.1}$$

The coordinates of the point P are

$$\begin{aligned} y &= r \sin \theta - t \sin(\theta + \varphi) \\ z &= -r \cos \theta + t \cos(\theta + \varphi) \end{aligned} \tag{H.2}$$

and, as usual,

$$dy/dz = \tan \psi = -\tan \tfrac{1}{2} (\theta_i + \theta + \varphi) \tag{H.3}$$

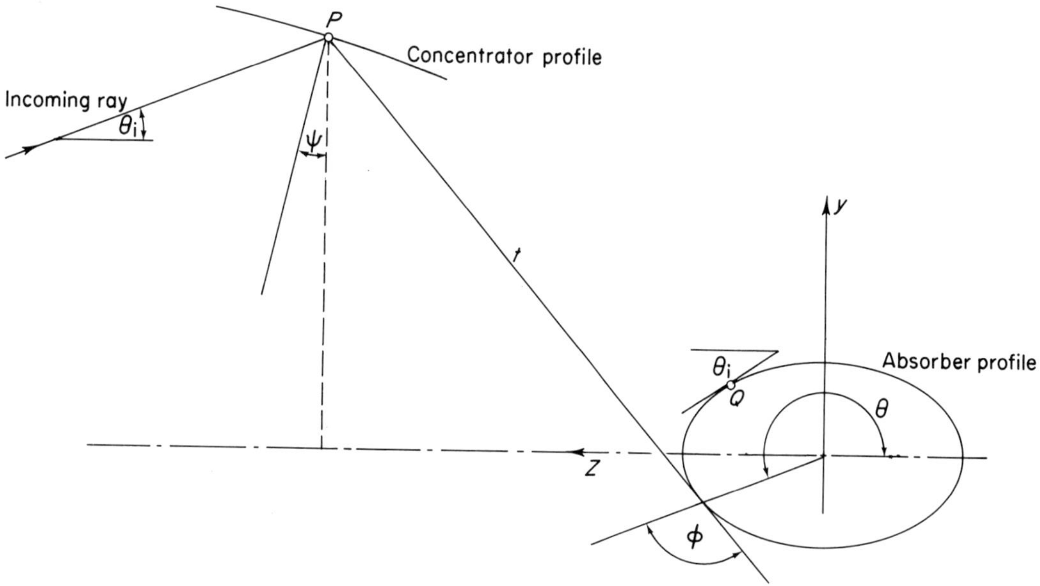

Fig. H.1 The differential equation for a 2D concentrator for a nonplane absorber.

Thus, on differentiating Eq. (H.2) and using Eq. (H.3), we obtain

$$\frac{\dot{r}\sin\theta + r\cos\theta - \dot{t}\sin(\theta+\varphi) - t\cos(\theta+\varphi)(1+\dot{\varphi})}{-\dot{r}\cos\theta + r\sin\theta + \dot{t}\cos(\theta+\varphi) - t\sin(\theta+\varphi)(1+\dot{\varphi})} + \tan\tfrac{1}{2}(\theta_i + \theta + \varphi) = 0 \tag{H.4}$$

In this equation t is the dependent variable and θ is the independent variable; r and φ are both known functions of θ. Equation (H.4) is therefore a first-order linear differential equation for the profile which could be solved by well-known methods.

For the involute portion of the profile we have the classical result

$$dt/d\theta = [r^2 + (dr/d\theta)^2]^{\frac{1}{2}} \tag{H.5}$$

for the current point.

We can, however, write down the solution to Eq. (H.4) in closed form using the "string method" described in Chapter 6. Thus, let s be the perimeter of the absorber profile measured from the point corresponding to $\theta = 0$; θ_0 is the polar angle of the point Q that separates the shadowed from the directly viewed portion of the absorber. Then, by simply imposing the condition of constant string length on the geometry of Fig. H.1, we obtain for the portion $\theta > \theta_0$

$$t = \frac{[s + s(\theta_0)] - r\cos(\theta - \theta_i) + r(\theta_0)\cos(\theta_0 - \theta_i)}{[1 - \cos(\theta + \varphi - \theta_i)]} \tag{H.6}$$

For the involute portion ($\theta < \theta_0$) we have

$$t = s \tag{H.7}$$

Rabl (1976a) set up these differential equations for the special case of great practical importance when the absorber cross section is circular. In this case $\dot{r} = 0$, $\varphi = \pi/2$, and Eq. (H.4) reduces to the form

$$(dt/d\theta) - t\tan[\theta - \tfrac{1}{2}(\theta_i + \theta + \pi/2)] - r = 0$$

or, setting $\frac{1}{2}(\theta - \theta_i + \pi/2) = \beta$

$$(dt/d\beta) - 2t\tan\beta = 2r \tag{H.8}$$

As Rabl showed, this can be integrated by making the substitution

$$t = u/\cos^2\beta$$

to give

$$t = \frac{r[(\theta - \theta_i - \pi/2) - \cos(\theta - \theta_i) + \text{const}}{1 + \sin(\theta - \theta_i)} \tag{H.9}$$

To determine the constant, we note that for $\theta = \theta_i + \pi/2$, we desire $t = r(\theta_i + \pi/2)$. This fixes the value of the constant in Eq. (H.9) and gives

$$t = \frac{r[(\theta + \theta_i + \pi/2) - \cos(\theta - \theta_i)]}{1 + \sin(\theta - \theta_i)} \tag{H.10}$$

for $\pi/2 + \theta_i \leq \theta \leq 3\pi/2 - \theta_i$. For $\theta \leq \theta_i + \pi/2$,

$$t = r\theta \text{ (involute)} \tag{H.11}$$

Again, we could have skipped over the differential equation and obtained the solution directly by the string method, noticing that for the circle $s = r\theta$, $\theta_0 = \theta_i + \pi/2$.

APPENDIX I

Deriving a Formula for Generalized Radiance

Of the several expressions which have been proposed for generalized radiance, we choose one of those proposed by Walther (1973) to illustrate the procedure. The expression, like all the others which have been suggested, is in the scalar wave model; here there is an immediate problem, the formulation of an energy flow vector analogous to the Poynting vector in electromagnetism, because the scalar wave amplitude, $u(\mathbf{r})$ is not a well-defined physical quantity in optics. Walther follows the formalism of quantum mechanics (see, e.g., Schiff, 1968) in assuming that the energy flow vector is, apart from scale factors,

$$\mathbf{P} = Im u^* \boldsymbol{\nabla} u \tag{I.1}$$

A derivation (which cannot be claimed to be rigorous) of this expression is as follows.

Maxwell's equations for a source-free vacuum are

$$\partial\mathbf{H}/\partial t = -\boldsymbol{\nabla} \times \mathbf{E}, \qquad \partial\mathbf{E}/\partial t = \boldsymbol{\nabla} \times \mathbf{H}$$

For a monochromatic field we have,

$$\partial\mathbf{H}/\partial t = -i\omega\mathbf{H}$$

so that, from the first of these, $\mathbf{H} = (-i/\omega)\boldsymbol{\nabla} \times \mathbf{E}$.

Thus from the second Maxwell equation, $\partial \mathbf{E}/\partial t =$

$$\nabla \times \{(-i/\omega)\nabla \times \mathbf{E}\} = (-i/\omega)\{\nabla(\nabla \cdot \mathbf{E}) - \nabla^2 \mathbf{E}\}$$

or, since $\nabla \cdot \mathbf{E} = 0$,

$$\partial \mathbf{E}/\partial t = (i/\omega)\nabla^2 \mathbf{E}$$

Now we make the scalar wave approximation, putting $\mathbf{E} \to u$ and we take for the light intensity

$$I = u^* u$$

The light energy in a volume V is

$$\int_V u^* u \, d^3 r$$

and its rate of change is

$$\begin{aligned}
\partial/\partial t \int_V u^* u \, d^3 r &= (-i/\omega) \int_V (u\nabla^2 u^* - u^* \nabla^2 u) \, d^3 r \\
&= (-i/\omega) \int_V \nabla \cdot (u \nabla u^* - u^* \nabla u) \, d^3 r \\
&= (-i/\omega) \int_A [u \nabla u^* - u^* \nabla u]_n \, dA
\end{aligned}$$

by Green's theorem, where $[\ ]_n$ denotes the outward normal component of the vector. We note that

$$u \nabla u^* - u^* \nabla u = 2iIm u \nabla u^*$$

Thus ignoring scaling and dimensionality we take the expression in Eq. (I.1) as the energy flow vector in the scalar wave field.

Walther then writes

$$\mathbf{P} = \mathrm{Re} u^*(\mathbf{r}) \int \mathbf{n} \bar{u}(\mathbf{n}) \exp(ik\mathbf{n} \cdot \mathbf{r}) \, d\Omega$$

where $\bar{u}(\mathbf{n})$ is defined by

$$u(\mathbf{r}) = \int \bar{u}(\mathbf{n}) \exp(ik\mathbf{n} \cdot \mathbf{r}) \, d\Omega$$

so that the energy flow through an element of area dA at $\mathbf{r}$ with normal

$\mathbf{n}_A$ is

$$d\phi = dA \int (\mathbf{n}_A \cdot \mathbf{n}) \mathrm{Re}\, u^*(\mathbf{r}) \bar{u}(\mathbf{n}) \exp(ik\mathbf{n} \cdot \mathbf{r})\, d\Omega \tag{I.2}$$

If the radiation is partially coherent, as it would be in almost all cases of interest in nonimaging optics, an ensemble average over the fields from all source points is taken. Eq. (I.2) can be compared with the geometrical optics equivalent in radiometry, which is

$$d\phi = dA \int (\mathbf{n}_A \cdot \mathbf{n}) B(\mathbf{r}, \mathbf{n})\, d\Omega \tag{I.3}$$

Where B is the radiance in direction $\mathbf{n}$ at the point $\mathbf{r}$. Consequently, we can identify the generalized radiance in Eq. (I.2) and, putting in the ensemble average signs, we have

$$B(\mathbf{r}, \mathbf{n}) = \mathrm{Re}\langle u^*(\mathbf{r}) \bar{u}(\mathbf{n}) \exp(ik\mathbf{n} \cdot \mathbf{r})\rangle \tag{I.4}$$

This is the expression given by Walther (1973).

APPENDIX J

Skew Rays in Hyperboloid Concentrator

We shall prove that any ray aimed at the rim of the entry aperture is reflected to some point on this rim. Let FF' in Fig. J.1 be the diameter of the entry aperture, so that F and F' are the foci of the hyperbolae in the plane of the diagram, and let P be a point on a hyperbola. The bisector PR of angle FPF' is tangent to the hyperbola and thus the tangent plane at P passes through PR. Any skew ray to the rim incident on the hyperbola at P passes through some point on the circle of which FF' is the diameter and which is in a plane perpendicular to the diagram, and all such rays generate an oblique circular cone. We now draw F_1F_1' through R to make the same angle with PR, as shown. A plane through F_1F_1' perpendicular to the diagram cuts the cone in a certain curve; the cone is a (degenerate) quadric surface and therefore

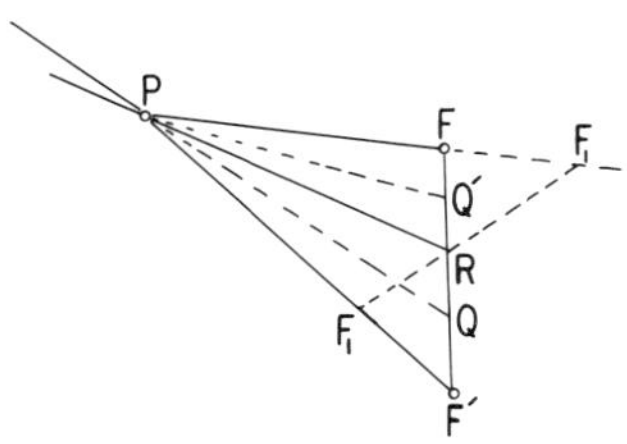

Fig. J.1 Geometry of hyperboloid concentrator for skew rays.

any plane intersects it in a conic section; but the plane through FF' intersects the cone in a circle by construction, and since the section through F_1F_1' has the same dimensions in and perpendicular to the plane of the diagram, it too must be a circle. An incident skew ray aimed at Q on the circle through FF' is reflected to some point Q' on the plane through FF', and by our construction this point is the mirror image in the tangent plane to the hyperboloid of the point in which the ray cuts the plane through F_1F_1'; thus Q' is on the circle through FF'.

APPENDIX K

Sine Relation for Hyperboloid/Lens Concentrator

With reference to Fig. K.1, the lens images the input beams inclined at angles $\pm\theta$ to the axis onto the edges FF' of the virtual source. Therefore, by Fermat's principle $[QA']+[A'F']=[AF']$, where the brackets denote optical path length, and similarly for the other beam. From the geometry of the hyperbola, $[AF']-[A'F']=2c$. Also $[QA']=2a \sin \theta$, since QA is normal to the beam. The desired result follows: $c=a \sin \theta$.

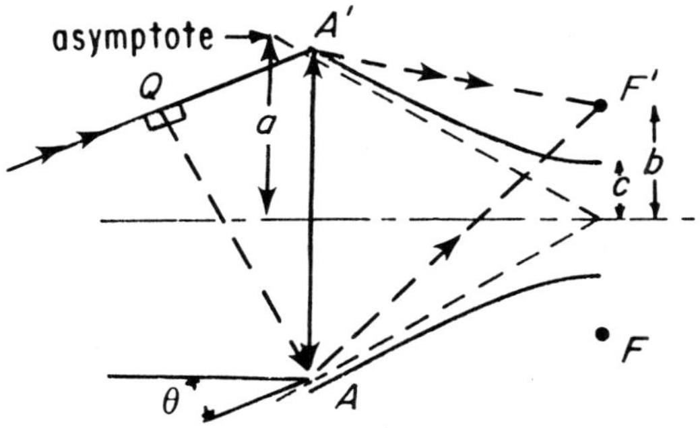

Fig. K.1 The thin lens at AA' images the input beam inclined at angle θ to the axis onto F'.

References

Baranov, V. K. (1965a). *Opt. Mekh. Prom.* **6**, 1-5.
A paper in Russian that introduces certain properties of CPCs.

Baranov, V. K. (1965b). "FOCON" (in Russian). Russian certificate of authorship 167327, specification published March 18, 1965.
Describes the 3D CPC, called a FOCON.

Baranov, V. K. (1966). *Geliotekhnika* **2**, 11-14 [*Eng transl.*: Parabolotoroidal mirrors as elements of solar energy concentrators. *Appl. Sol. Energy* **2**, 9-12.].

Baranov, V. K. (1967). "Device for Restricting in One Plane the Angular Aperture of a Pencil of Rays from a Light Source" (in Russian). Russian certificate of authorship 200530, specification published October 31, 1967.
Describes certain illumination properties of the 2D CPC, called in other Russian publications a FOCLIN.

Baranov, V. K., and Melnikov, G. K. (1966). Study of the illumination characteristics of hollow focons. *Sov. J. Opt. Technol.* **33**, 408-411.
Brief description of the principle, with photographs to show illumination cut-off and transmission-angle curves plotted from measurements.

Bassett, I. M. (1984). Limits to concentration in resonant cavities. *Opt. Acta.* **31**, 1211-1232.

Bassett, I. M. (1985). Limits to concentration by passive means. *Opt. Acta* **32**, 1577-1592.

Bassett, I. M. (1986). Limits to concentration by focussing. *Opt. Acta* **33**, 279-286.

Bassett, I. M., and Derrick, G. H. (1978). The collection of diffuse light onto an extended absorber. *Opt. Quantum Electron.* **10**, 61-82.

Bassett, I. M., and Winston, R. (1984). Limits to concentration in physical optics and wave mechanics. *Opt. Acta* **31**, 499-505.

Baylor, D. A., and Fettiplace, R. (1975). Light and photon capture in turtle receptors. *J. Physiol. (London)* **248**, 433-464.

Born, M., and Wolf, E. (1975). "Principles of Optics," 5th Ed. Pergamon, Oxford, England.

Burkhard, D. G., and Shealy, D. L. (1975). Design of reflectors which will distribute sunlight in a specified manner. *Sol. Energy* **17**, 221–227.

Butler, B. L., and Pettit, R. B. (1977). Mirror materials and selective coatings. ERDA Semiannu. Rev. Rep. SAND 77-0111. Sandia Laboratories. Albuquerque, New Mexico.

Carter, W. H., and Wolf, E. (1977). Coherence and radiometry with quasihomogeneous planar sources. *J. Opt. Soc. Am.* **67**, 785–796.

Collares-Pereira, M., Rabl, A., and Winston, R. (1977). Lens-mirror combinations with maximal concentration. *Appl. Opt.* **16**, 2677–2683.

Cornbleet, S. (1976). "Microwave Optics." Academic Press, New York.

Duffie, J. A., and Beckman, W. A. (1980). "Solar Engineering of Thermal Processes." Wiley, New York.

Eiden, R. (1968). Calculations and measurements of the spectral radiance of the solar aureole. *Tellus* **20**, 380–399.

Elmer, W. B. (1980). "The Optical Design of Reflectors," 2nd Ed. Wiley, New York.

Ford, G. (1977). Personal communication. Energy Design Corp., Memphis, Tennessee.

Gabor, D. (1964). Light and information. *Prog. Opt.* **1**, 112–152.

Garrison, J. D. (1979a). Optimization of a fixed solar thermal collector, *Sol. Energy* **23**, 93–102.

Garrison, J. D. (1979b). Radiation collection by an optimized fixed collector. *Sol. Energy* **23**, 103–109.

Garwin, R. L. (1952). The design of liquid scintillation cells. *Rev. Sci. Instrum.* **23**, 755–757.

Gleckman, P. (1988). Achievement of ultra high solar concentration with potential for efficient laser pumping. *Appl. Opt.* **27**, 4385–4391.

Gleckman, P., O'Gallagher, J. and Winston, R. (1989). Concentration of sunlight to solar-surface levels using non-imaging optics. *Nature* **339**, 198–200.

Gordon, J. (1988). Personal communication.

Grum, F., and Becherer, R. J. (1979). "Radiometry" (Vol. 1 of "Optical Radiation Measurements"). Academic Press, New York.

Harper, D. A., Hildebrand, R. H., Pernic, R., and Platt, S. R. (1976). Heat trap: An optimized far infrared field optics system. *Appl. Opt.* **15**, 53–60.

Hildebrand, R. H., and Winston, R. (1982). Throughput of diffraction-limited field optics system for infrared and millimetric telescope. *Appl. Optics* **21**, 1844–1846. See also errata in *Appl. Opt.* **24**, 616.

Hildebrand, R. H., Whitcomb, S. E., Winston, R., Stiening, R., Harper, D. A., and Moseley, S. H. (1977). Submillimeter photometry of extragalactic objects. *Astrophys. J.* **216**, 698–705.

Hildebrand, R. H., Whitcomb, S. E., Winston, R., Stiening, R., Harper, D. A., Moseley, S. H. (1978). Submillimeter observations of the galactic center. *Astrophys. J.* **219**, L101–L104.

Hinterberger, H., and Winston. R. (1966a). Efficient light coupler for threshold Čerenkov counters. *Rev. Sci. Instrum.* **37**, 1094–1095.

Hinterberger, H., and Winston, R. (1966b). Gas Čerenkov counter with optimized light-collecting efficiency. *Proc. int. Conf. Instrum. High Energy Phys.* 205–206.

Hinterberger, H., and Winston, R. (1968a). Efficient design for lucite light pipes coupled to photomultipliers. *Rev. Sci. Instrum.* **39**, 419–420.

Hinterberger, H., and Winston, R. (1968b). Use of a solid light funnel to increase phototube aperture without restricting angular acceptance. *Rev. Sci. Instrum.* **39**, 1217–1218.

Hinterberger, H., Lavoie, L., Nelson, B., Sumner, R. L., Watson, J. M., Winston, R., and Wolfe, D. M. (1970). The design and performance of a gas Čerenkov counter with large phase space acceptance *Rev. Sci. Instrum.* **41**, 413–418.

Holter, M. L., Nudelman, S., Suits, G. H., Wolfe, W. L., and Zissis, G. J. (1962). "Fundamentals of Infrared Technology." Macmillan, New York.

Hottel, H. (1954). Radiant heat transmission. *In* "Heat Transmission" (W. H. McAdams, ed.), 3rd Ed. McGraw-Hill, New York.

Keene, J., Hildebrand, R. H., Whitcomb, S. E., and Winston, R. (1978). Compact infrared heat trap field optics. *Appl. Opt.* **17**, 1107–1109.

Kreider, J. F., and Kreith, F. (1981). "McGraw-Hill Solar Energy hand book." McGraw-Hill, New York.

Krenz, J. H. (1976). "Energy Conversion and Utilization." Allyn & Bacon, Rockleigh, New Jersey.

Levi-Setti, R., Park, D. A., and Winston, R. (1975). The corneal cones of *Limulus* as optimized light collectors. *Nature (London)* **253**, 115–116.

Luneburg, R. K. (1964) "Mathematical Theory of Optics." Univ. of California Press, Berkeley. This material was originally published in 1944 as loose sheets of mimeographed notes and the book is a word-for-word transcription.

McIntire, W. (1980). New reflector design which avoids losses through gaps between tubular absorber and reflector. *Sol. Energy.* **25**, 215–220.

McIntire, W., and Winston, R. (1981). Design considerations for reducing optical losses due to gaps between absorbers and their reflectors (B. H. Glenn and G. E. Franta, eds) pp. 211–273. Proc. 1981 Ann Meeting American Section of the Int. Solar Energy Society.

McKechnie, T. S. (1975). Chapter 4 of "Laser Speckle and Related Phenomena" (J. C. Dainty, ed.). Springer-Verlag, Heidelberg.

Marcuse, D. (1972). "Light Transmission Optics." Van Nostrand-Reinhold, Princeton, New Jersey.

Maxwell, J. C. (1958). On the general laws of optical instruments. *Q. J. Pure Appl. Math.* **2**, 233–247.

Morgan, S. P. (1958). General solution of the Luneburg lens problem *J. Appl. Phys.* **29**, 1358–1368.

Ning, N., (1988). Three-dimensional ideal θ_2/θ_2 angular transformer and its uses in fiber optics, *Appl. Opt.* **27**, 4126–4130.

Ning, X, Winston, R., and O'Gallagher, J. (1987a). Dielectric totally internally reflecting concentrators, *Appl. Opt.* **26**, 300–305.

Ning, X., O'Gallagher, J., and Winston, R., (1987b). Optics of two-stage photovoltaic concentrators with dielectric second stages, *Appl. Opt.* **26**, 1207–1212.

O'Gallagher, J., and Winston, R. (1983). Development of compound parabolic concentrators for solar energy. *Int. J. Ambient Energy* 171–186.

O'Gallagher, J. J., and Winston, R. (1986). Test of a "trumpet" secondary concentrator with a paraboloidal dish primary. *Sol. Energy* **36**, 37.

O'Gallagher, J., Winston, R., Suresh, D., and Brown, C. T., (1987) Design and test of an optimized secondary concentrator with potential cost benefits for solar energy conversion. *Energy* **12**, 217–226.

Ortobasi, U. (1974). "Proposal to Develop an Evacuated Tubular Solar Collector Utilizing a Heat Pipe." Proposal to National Science Foundation (unpublished).

Pettit, R. B. (1977). Characterization of the reflected beam profile of solar mirror materials. *Sol. Energy* **19**, 733–741.

Ploke, M. (1967). Lichtführungseinrichtungen mit starker Konzentrationswirkung. *Optik* **25**, 31–43.

Ploke, M. (1969). "Axially Symmetrical Light Guide Arrangement. "German patent application No. 14722679.

Rabl, A. (1976a). Solar concentrators with maximal concentration for cylindrical absorbers. *Appl. Opt.* **15**, 1871–1873.

Rabl, A. (1976b). Comparison of solar energy concentrators. *Sol. Energy* **18**, 93–111.

Rabl, A. (1976c). Optical and thermal properties of compound parabolic concentrators. *Sol. Energy* **18**, 497–511.

Rabl, A., Goodman, N. B., and Winston, R. (1979). Practical design considerations for CPC solar collectors. *Solar Energy* **22**, 373–381.

Rabl, A. (1985). "Active Solar Collectors and Their Applications." Oxford Univ. Press, New York.

Rabl, A., and Winston, R. (1976). Ideal concentrators for finite sorces and restricted exit angles. *Appl. Opt.* **15**, 2880–2883.

Russell, J. L. (1976). Principles of the fixed mirror solar concentrator. *Proc. Soc. Photo-optical Instru. Eng.* **85**, 139–145.

Sato, S., Hayakawa, S., Lange, A. E., Matsumoto, T., Matsuo, H., Murakami, H., and Richards, P. L. (1987). Horn antenna with low sidelobe response for observations of diffuse celestial radiation. *Appl. Opt.* **26**, 410–415.

Scharlack, R. S. (1977). All-dielectric compound parabolic concentrator. *Appl. Opt.* **16**, 2601–2602.

Schiff, L. I. (1968). "Quantum Mechanics," 3rd Ed. McGraw-Hill, New York.

Smith, W. J. (1966). "Modern Optical Engineering," McGraw-Hill, New York.

Spectrolab, Inc. (1972). "Investigation of Terrestrial Photovoltaic Power Systems with Sunlight Concentration," Final Rep. Spectrolab, Inc., Sylmar, California.

Sparrow, E. M., and Cess R. D. (1978). Radiation Heat Transfer," augmented edition, McGraw Hill, New York.

Stamnes, J. J. (1986). "Waves in Focal Regions." Hilger, Bristol, England.

Stavroudis, O. N. (1973). Comments on Archimedes' burning glass. *Appl. Opt.* **12**, A16.

Walther, A. (1967). Gabor's theorem and energy transfer through lenses. *J. Opt. Soc. Am.* **57**, 639–644.

Walther, A. (1968). Radiometry and coherence. *J. Opt. Soc. Am.* **58**, 1256–1259.

Walther, A. (1973). Radiometry and coherence. *J. Opt. Soc. Am.* **63**, 1622–1623.

Weatherburn, C. E. (1931). "Differential Geometry of Three Dimensions." Cambridge Univ. Press, London.

Welford, W. T. (1974). "Aberrations of the Symmetrical Optical System." Academic Press, New York.

Welford, W. T. (1977). Optical estimation of statistics of surface roughness from light scattering measurements. *Opt. Quantum Electron.* **9**, 269–287.

Welford, W. T. (1986). "Aberrations of Optical Systems." Hilger, Bristol, England.

Welford, W. T., and Winston, R. (1982a). Upper bound on the efficiency of certain nonimaging concentrators in the physical optics model. *J. Opt. Soc. Am.* **72**, 1244-1248.

Welford, W., and Winston, R. (1987). Generalized radiance and practical radiometry. *J. Opt. Soc. Am.* **A4**, 545-547.

Welford, W. T., O'Gallagher, J., and Winston, R., (1987). Axially symmetric nonimaging flux concentrators with the maximum theoretical concentration ratio, *J. Opt. Soc. Am. A* **4**, 66-68.

Williamson, D. E. (1952). Cone channel condenser optics. *J. Opt. Soc. Am.* **42**, 712-715.

Winston, R. (1970). Light collection within the framework of geometrical optics. *J. Opt. Soc. Am.* **60**, 245-247.

Winston, R. (1974). Principles of solar concentrators of a novel design. *Sol. Energy* **16**, 89-95.

Winston, R. (1975). Development of the compound parabolic collector. *Proc. Soc. Photo-opt. Instrum. Eng.* **68**, 136-144.

Winston, R. (1976a). Dielectric compound parabolic concentrators. *Appl. Opt* **15**, 291-292.

Winston, R. (1976b). U.S. letters patent 3923 381, "Radiant Energy Concentration."

Winston, R. (1976c). U.S. letters patent 3957 031, "Light Collectors in Cylindrical Geometry."

Winston, R. (1977a). U.S. letters patent 4003 638, "Radiant Energy Concentration."

Winston, R. (1977b). U.S. letters patent 4002 499, "Cylindrical Concentrators for Solar Energy."

Winston, R. (1978a). Cone collectors for finite sources. *Appl. Opt.* **17**, 688-689.

Winston, R. (1978b). Ideal light concentrators with reflector gaps. *Appl. Opt.* **17**, 1668.

Winston, R. (1980). Cavity enhancement by controlled directional scattering. *Appl. Opt.* **19**, 195-196.

Winston, R., and Enoch, J. M. (1971). Retinal cone receptor as an ideal light collector. *J. Opt. Soc. Am.* **61**, 1120-1121.

Winston, R., and Hinterberger, H. (1975). Principles of cylindrical concentrators for solar energy. *Sol. Energy* **17**, 255-258.

Winston, R., and Ning, X. (1988). Generalized radiance of uniform Lambertian sources. *J. Opt. Soc. Am.* **A4**, 516-519.

Winston, R., and Welford, W. T. (1979a). Geometrical vector flux and some new nonimaging concentrators, *J. Opt. Soc. Am.* **69**, 532-536.

Winston, R. and Welford, W. T. (1979b). Ideal flux concentrators as shapes that do not disturb the geometrical vector flux field: A new derivation of the compound parabolic concentrator, *J. Opt. Soc. Am.* **69**, 536-539.

Winston, R., and Welford, W. T. (1980). Design of nonimaging concentrators as second stages in tandem with image-forming first stage concentrators. *Appl. Opt.* **19**, 347-351.

Winston, R., and Welford, W. T. (1982a). Efficiency of nonimaging concentrators in the physical optics model. *J. Opt. Soc. Am.* **72**, 1564-1566.

Winston, R., and Welford, W. T. (1982b). *J. Opt. Soc. Amer.* **72**, 1564-1566.

Witte, W. (1965). Cone channel optics. *Infrared Phys.* **5**, 179-185.

Wolf, E. (1978). Coherence and radiometry, *J. Opt. Soc. Am.* **68**, 6-17.

Index

L

M

N

O

P

Q

R

S